STANDARDS on lowland farms

Rer

Peter Lack

British Trust for Ornithology

JOINT NATURE CONSERVATION COMMITTEE

MAFF Ministry of Agriculture Fisheries and Food

London: HMSO

*All illustrations by Steve Carter except for Robin
(p. 17) by Steve Cull, Redwings (p. 24) by Kevin
Baker and Kingfisher (p. 112) by Thelma Sykes*

*Cover photographs by: Geoff du Feu/Planet Earth
Pictures (main picture) and Roger Tidman (inset
picture)*

Printed in the United Kingdom for HMSO
Dd294724 10/92 C20 G531 10170

TABLE OF CONTENTS

BOXES

For nearly thirty years the British Trust for Ornithology has been monitoring the status of the common birds breeding in farmland in Britain. In that time the numbers of some have increased, numbers of others have remained much the same, and of others decreased. In a few cases the changes have been substantial, and it is the decreases in particular which have caused some concern among conservationists because of the possibility of any more general effects of farming on the quality of the environment. Farmers share these concerns, and wish to maximise the beneficial effects and minimise any adverse effects of their activities, while continuing to make a living from the land.

Until fairly recently environmental organisations directed most of their resources towards studying natural and semi-natural habitats. They did not ignore farmland – for example the Nature Conservancy Council funded the British Trust for Ornithology's Common Birds Census for twenty five years (now continued by the Joint Nature Conservation Committee) – but they largely left the management of farmland to the farmers. This situation is now changing. There is a growing awareness that all areas and land uses are interrelated, and that farmland cannot be isolated. Everybody wants good quality food at affordable prices and a good quality environment, and there need not be a conflict between the priorities of agriculture and other activities.

This growing interest in farmland has stimulated farming organisations and others into offering advice on environmental matters. This is being promoted both by publishing leaflets and advisory booklets, and in making advice available to individual farmers on a personal basis. Advisers with specialist knowledge, employed by for example ADAS of the Ministry of Agriculture Fisheries and Food (MAFF) and the Farming and Wildlife Trust (FWT), are increasingly being asked to visit farms and discuss specific individual problems on the ground.

The advisory material currently available describes a variety of conservation measures and management practices which will benefit conservation, but rarely explains the reasons for the advice offered, nor does it specify in detail the likely consequences of different practices on particular species or groups of wildlife. This book is designed specifically to fill this need for birds. The primary aims are therefore:

1 to describe what birds need in the farm landscape, and how they use the various habitat features;

2 to indicate which are the most important factors in determining the distribution and abundance of birds;

3 to point out the effects on birds of different management practices; and

4 to show how management practices might be adapted to improve the habitat for birds, while remaining consistent with maintaining the farm as an economically viable enterprise.

The distribution and abundance of birds and the effects of management practices are quantified wherever possible. Readers can therefore gauge for themselves the benefits and drawbacks of different habitat features and management practices.

The primary emphasis is on maintaining and managing features of interest which are already present. This is usually the easiest, cheapest and most efficient way of benefitting birds. Such features will often already contain many of the important characteristics. It is also often the only way of conserving plants and less mobile animals. However, sometimes it is practical to create new areas of wildlife habitat on farms, and such creation is covered explicitly where it is feasible, although in all cases when it is contemplated, it is essential to ensure that existing features of interest are not destroyed in the process. Finally, the book considers some of the alternative enterprises and scenarios for farmland management, which are being implemented now or proposed for the future, for example set-aside, extensification and farm forestry.

The first chapter describes the scope of the book. It gives a brief resume of the major sources of information, with a list of the common bird species found in British farmland. These are the birds which are the main subjects of the book and for which the suggestions are primarily aimed. Birds are tabulated under habitat categories and any regional patterns in their distribution are indicated.

Chapter two introduces some general principles of bird conservation in farmland and should be read as a background to the more specific suggestions made in later chapters. It includes a short discussion of regional differences, both of the birds and of farming. These differences have some bearing on what

follows as not all birds occur in all areas and some farming practices are localised.

Chapters three to eleven form the practical section of the book. Each of the major habitat features to be found in farmland is considered in turn. For each, the important aspects for birds are discussed, and how the birds use them and are influenced by them. Where possible specific examples are included with illustrations, many of which are presented as self-contained 'boxes'. Particular emphasis is placed on the management options and their effects on birds. At the end of each chapter there is a summary of the main points and the effects of different management practices.

The final chapter presents a general summary of the main messages of earlier chapters, points to some of the aspects which as yet we know rather little about, and to some of the farming initiatives that have been proposed. The emphasis is on how such initiatives might be of greatest help to birds in general, especially those species which are at present declining in farmland.

ACKNOWLEDGEMENTS

The book arises through a collaboration between the British Trust for Ornithology (BTO), Joint Nature Conservation Committee (JNCC) and the Ministry of Agriculture, Fisheries and Food (MAFF). The main data source is the Common Birds Census. This is a continuing BTO project which has been funded by the Nature Conservancy Council and now by the JNCC. In 1985 MAFF contracted the BTO to examine the habitat requirements of farmland birds using the CBC database, and to produce this book as a practical advisory aid for farmland managers.

Throughout, the project has been overseen by an Advisory Committee consisting of representatives of the BTO, JNCC and MAFF, under the chairmanship of Michael Way (until February 1988) and Peter Greig-Smith (after February 1988). Members of this group during part or all of the project were Stuart Dorward (BTO), Rob Fuller (BTO), Colin Galbraith (JNCC), Tony Hardy (MAFF), Eric Hunter (MAFF), Raymond O'Connor (BTO), Philip Oswald (JNCC), and Mike Pienkowski (JNCC). Further contributions were made by Jeremy Greenwood (BTO), John Byng (MAFF) and Mike Moser (BTO), and Nigel Allen, Gordon Turnbull and Sian Jones (all MAFF) were Minuting Secretaries in turn.

The book has benefited considerably from the comments of most of these people on earlier drafts and I must single out Peter Greig-Smith, as Chairman of the Committee when the book was being written. He provided many detailed comments and suggestions at various stages and also handled all the negotiations with HMSO over the publishing.

The text was improved greatly by the editorial pen of Robert Hudson, the diagrams were all kindly prepared by Liz Murray and the index was compiled by Su Gough. Most of the illustrations are by Steve Carter and he also arranged for others by Kevin Baker, Steve Cull and Thelma Sykes.

The following kindly gave permission to reproduce diagrams: Cambridge University Press (those in Boxes 3.3, 3.6 and 3.9), British Crop Protection Council (in Box 3.5), Elsevier Applied Science Publishers (in Box 8.7) and Gauthier-Villars (in Box 8.6).

The Common Birds Census relies on volunteer birdwatchers to do the census work on which so many results and analyses are based. Many have contributed essential information to much of the detail described in this book, and without them the work would not have been possible.

To all of these people and organisations I am very grateful.

CHAPTER 1

Scope and Sources

1.1   Scope of the book

The birds considered here are those which occur naturally in farmland. Most of these species are of no particular economic interest to the farmer, although game birds can be directly and financially beneficial through sport-shooting, and a few others can be classed as pests. But because most birds eat insects and other potential pests, the majority are, on balance, more likely to be beneficial than a nuisance to the farmer, although their effect may be small. There is, however, great aesthetic and environmental interest in birds.

One of the main aims of this book is to identify management practices which benefit birds. This may sometimes result in a conflict between the management practice which is of most benefit to birds and that which is the most efficient and beneficial to a farmer. This applies especially to chapter 6 on fields and crops. Such potential conflicts are not ignored. Usually the ideal management for birds is presented first, but if this is not very practical for efficient farming it is followed by some suggestions for realistic compromises, which will often still produce considerable benefits for birds.

The common breeding and wintering bird species being considered are given in Box 1.1. They are listed in groups firstly by their preference for a main habitat feature and secondly by regional distribution. Many of them are widespread in farmland over all of Britain, and some also in other habitat types, but some are more restricted in range, either geographically or to one or more particular habitat features.

The book is designed primarily as a guide to improving farmland for birds. However many other animals and plants also depend on farmland. Many of the conservation suggestions made are beneficial to these others too, but it is beyond the scope of this book to consider them in detail. A few suggestions may be detrimental to other wildlife. Such cases are discussed individually as also are a few cases where there is a well established conservation message appropriate for another wildlife group but which has little or no effect on birds.

There are several advantages to using birds in such a book. Farmland holds substantial populations of many bird species[1], even if it is not their most preferred habitat. There are few farmland species that are nationally rare, but for example farmland is the main breeding habitat of the Stone Curlew. Some other breeding waders, e.g. Snipe, Redshank and Lapwing, and seed eating species such as Linnet, Tree Sparrow, and Corn Bunting also depend heavily on farmland and numbers have been declining steadily in recent years. In winter the farmland populations of some species, even quite common ones such as Lapwing, Golden Plover, Fieldfare and Redwing, are very important in international conservation terms[2].

Birds are also good indicators of the general health of the environment, especially of pollutants such as persistent pesticides (e.g. ref 3). Birds have been relatively well studied and the effects of farming practices on their populations are rather better known than for many groups. The effects of conservation measures can therefore be predicted more precisely, and, as noted above, such measures are often beneficial to other wildlife groups as well. Finally, birds have considerable popular appeal and farmland and the countryside generally are places where farmers and the public wish to see them thrive.

The main emphasis of the book is on lowland farms below about 300 metres above sea level, although this is not a fixed limit. This somewhat arbitrary level has been imposed largely to exclude purely upland areas, including hill farming, unenclosed sheep walk, and moorland. These are certainly farmland but are rather different in habitat structure and farming regimes to those in the lowlands. Upland farms are also often subject to a rather different economic structure because of particular subsidies and grants.

The book is concerned mainly with agriculture and farmland in Britain, and the emphasis on lowland farms means that it is less relevant to parts of the north and west of the country than to the south and east. However, farmland in Britain is similar in many respects to that in other nearby countries, both in terms of its physical habitat structure and in its economic structure because of the Common Agricultural Policy of the European Economic Community. Therefore most points are relevant to neighbouring European countries, and indeed some of the examples are from these countries. Some points are relevant in general terms to other parts of the world as well.

The situations described are those which exist in the late 1980s and early 1990s. Historical aspects are not covered in any detail although they are mentioned where necessary. A major analysis of the effects of farming on birds since the early 1960s, mainly using BTO data, was given by Raymond O'Connor and Michael Shrubb in their book *Farming and Birds*[1]. Some predictions and suggestions for the future are given where relevant.

1.2   Sources of Information

BTO projects

The major source of information is the Common Birds Census (CBC), a territory-mapping project run by the British Trust for Ornithology (BTO) under a contract from the Joint Nature Conservation Committee (JNCC). Some of the results quoted here have been published previously but other information comes directly from BTO files. The CBC started in 1962, at the request of the Nature Conservancy, as a response to the need for quantified information on what appeared to be major declines in the populations of many bird species breeding in farmland caused by organochlorine insecticides, such as dieldrin and DDT. It was soon apparent that the CBC was a useful general monitor of breeding bird population sizes and changes in farmland. In 1964 the survey was extended into woodland and it has continued every year since then. The trends and changes in population sizes over twenty seven years of all the bird species which are monitored by the project have recently been published[4].

Full details of the methods used for the project are given elsewhere[4,5]. A volunteer observer visits a plot of land, which in farmland averages about 70 ha, eight to ten times during the breeding season (April to early July). The position of all birds seen or heard on each visit is marked onto a large scale map (1:2500) using different codes for each species and activity. At the end of the season all records for each species are collated onto species maps, which are then analysed by a trained professional to estimate the number of breeding territories present on that plot of land in that year. Comparing the numbers between one year and the previous one for all plots censused in both years, a measure of change can be obtained.

In addition to their bird records the observers also provide a detailed map of habitat features on the plot, including notes on the crops growing in each field in each year. The positions of bird territories can therefore be compared to the habitat features and land usage enabling analyses of the birds' requirements.

In all, about 100 plots of farmland and about eighty plots of woodland are censused each year. On average each plot is censused for seven years, and there are now about 3000 'plot-years' of information in the BTO files for farmland. Plots are chosen by the volunteer observers, but they are asked to choose a plot which is more or less typical of the surrounding area. To confirm that this is indeed the case, the types of farmland on CBC plots have been compared to the general distribution of farming types in the areas as shown by the annual agricultural statistics compiled by MAFF and the Department of Agriculture and Fisheries in Scotland (DAFS)[6]. The authors of this analysis found that the CBC plots have been fairly representative of lowland England during the whole CBC period, although there are more differences in upland areas. Therefore, any population changes or factors operating on CBC plots are representative of those occurring in lowland England, but more caution is needed in the uplands and in the north and west of Britain.

Results from some other BTO projects, some wholly or partly funded by JNCC, have also been incorporated into the book where appropriate. The Waterways Bird Survey started in 1974 and is a similar type of survey to the Common Birds Census but along rivers and canals. The Nest Record Scheme has been running since 1939 and is designed to provide information on the nest sites and breeding success of birds in Britain. A total of about 30,000 cards is submitted to the scheme each year, 25–30% of which are from farmland. Other BTO projects used include distribution atlases of birds in Britain and Ireland in the breeding season[7] and the winter[2], and several individual surveys of particular species or groups, such as a survey of breeding waders of wet meadows in 1982[8], a survey of breeding Lapwings in 1987[9], and a survey of birds in hedgerows in winter[10].

Other sources

The other major source of information is the published literature, both scientific and popular. The latter often contains much useful general information, though unfortunately it is often not specific as to the origin of details.

Other organisations and individuals have also been studying birds and farmland. Some of the results have been published but some have only appeared in unpublished reports. These have been used with permission where required. The Royal Society for the Protection of Birds (RSPB)

has worked especially on the more natural habitats within farmland, such as woods[11] or rivers, and they have worked extensively on the Stone Curlew[12]. The Game Conservancy Trust has been studying the ecology of the Grey Partridge for many years, e.g. ref 13. Their work as part of the Cereals and Gamebirds Research Project, and what are now called 'conservation headlands' (see ch 5), is especially relevant, although there is as yet only limited information on birds other than game birds. MAFF has carried out environmental work on their Experimental Husbandry Farms. The most relevant here is the Boxworth Project which has, amongst other things, been considering the effects of different pesticide regimes on wildlife, including birds[14,15]. The Joint Nature Conservation Committee, the government agency with responsibility for conservation at the national and international level, as well as funding several other organisations including the BTO, also carries out research work itself such as a project on lowland farmland in southern Scotland[16], and a current major programme on low intensity farmland particularly in agriculturally less favoured areas[17]. The Wildfowl and Wetlands Trust has done work in respect of geese and other wildfowl and their use of different crops and field types for grazing (see e.g. ref 18).

Other projects that have included work on birds in farmland, are the Countryside Commission's Demonstration Farms Project[19], and work by the Institute of Terrestrial Ecology especially on hedges[20] and on pesticides[21].

Several organisations produce advisory leaflets and booklets. These include ADAS, an executive agency of MAFF, the JNCC, the Game Conservancy Trust, the British Association for Shooting and Conservation, the RSPB, the British Trust for Conservation Volunteers and the Farming and Wildlife Trust with their associated county groups. Addresses of these and other relevant organisations with interests in farming and the countryside are in Appendix 2.

Box 1.1 The common birds in British farmland

The following tables list the species which are considered here to be 'common' birds in British farmland in the breeding season, in winter or both.

For the breeding season, fifty-five species are included. These are the forty nine species for which a Common Birds Census index of population level has been calculated specifically for farmland in the late 1980s[4], five species for which an index is calculated similarly over all habitats[4] but which are predominantly farmland birds, and the Rook. An index is not calculated for this species only because its colonial nesting habits make it unsuited to censusing using the CBC. (An index is calculated if the species was recorded holding at least one territory on at least fifteen plots in a year.)

Twelve of these fifty five species are summer visitors to Britain, spending the winter south of the Sahara in Africa or, in a few cases, around the Mediterranean. The remaining forty three species occur in farmland in the winter as well as in the breeding season although the numbers of some increase or decrease considerably compared to the summer. A proportion of the population of some species moves out of farmland in the winter either to other habitats or out of Britain altogether. The numbers of some other species increase and may be boosted by immigrants from northern or eastern Europe. In addition six species which do not breed in lowland farmland in Britain are considered to be 'common' in farmland in winter. Some other species, for example various goose species, could also be added to this last list because of their economic effects, but the list is restricted to those which are fairly widespread over farmland generally rather than to a few particular sites.

The species are listed both under their main habitat preferences (Tables 1: summer, and 2: winter), and by their regional distribution (Tables 3: summer, and 4: winter). Most will occur sometimes either in other habitats or in other regions.

Sources: mainly Marchant et al.[4], Sharrock[7], Lack[2].

Table 1
Habitat distribution (Breeding season)
Each of the fifty five common breeding species is placed into one of the six 'Nesting and Feeding' categories. Those which are purely summer visitors are marked with an asterisk.

a Nesting and feeding in the fields
Lapwing, Skylark, Yellow Wagtail.

b Nesting and feeding mainly in hedges or other field boundaries

(these are species which need some woody vegetation in their nesting territory, but only occur in the edges of woodland – a woodland index is not calculated)
Kestrel, Red-legged Partridge, Grey Partridge, Little Owl, Lesser Whitethroat*, Goldfinch, Corn Bunting.

c Nesting and feeding in hedges and woods

(these are species for which both a farmland and a woodland index are calculated)
Woodpigeon, Wren, Dunnock, Robin, Blackbird, Song Thrush, Mistle Thrush, Blackcap*, Whitethroat*, Willow Warbler*, Long-tailed Tit, Blue Tit, Great Tit, Magpie, Carrion Crow, Tree Sparrow, Chaffinch, Greenfinch, Linnet, Yellowhammer.

d Nesting and feeding mainly in woods or scrub

(these species are not typical of hedges, although they may use them especially when population levels are high)
Pheasant, Stock Dove, Turtle Dove*, Cuckoo*, Green Woodpecker, Great Spotted Woodpecker, Garden Warbler*, Chiffchaff*, Goldcrest, Spotted Flycatcher*, Coal Tit, Treecreeper, Jay, Rook, Jackdaw, Bullfinch.

e Nesting and feeding mainly by ponds or streams

Mallard, Moorhen, Pied Wagtail, Sedge Warbler*, Reed Bunting.

f Nesting and feeding mainly near farm buildings

Collared Dove, Swallow*, Starling, House Sparrow.

The following species, all of which are included in one of the above categories, also **feed extensively in (or while flying over) the fields**

Kestrel, Red-legged Partridge, Grey Partridge, Pheasant, Moorhen, Stock Dove, Woodpigeon, Swallow, Mistle Thrush, Magpie, Jackdaw, Rook, Carrion Crow, Starling.

Several other species often **feed near the edge of fields or in small paddocks**, including *Blackbird, Song Thrush* and several *finches* and *buntings.*

Table 2
Habitat distribution (Winter)

The six extra species:

Feeding mainly in the fields
Golden Plover, Snipe, Black-headed Gull, Meadow Pipit.

Feeding in fields and hedges
(and more rarely in woods)
Fieldfare, Redwing.

The resident species:

All the species in Table 1 which stay through the winter also remain to feed in the same habitat features then. Feeding in the fields is more common in the winter.

Table 3
Regional distribution (Breeding season)
The following are indications of where a species is most common. Information is derived from the density of the species on farmland Common Birds Census plots. The categories are necessarily slightly arbitrary, and each species is placed in only one of them. The main restrictions of range are noted, but in addition many species are scarcer on higher ground, especially in Scotland, and several more are absent from the northern and western islands. For precise distribution maps for Britain and Ireland in the breeding season, see Sharrock[7].

Commonest in South and East

Mallard, Stock Dove (absent from parts of Scotland), *Collared Dove, Dunnock, Song Thrush, Sedge Warbler, House Sparrow, Tree Sparrow* (absent from parts of Scotland, Wales and SW England), *Greenfinch, Linnet.*

Commonest in South and West

Little Owl (absent Scotland and higher ground in west), *Green Woodpecker* (absent from N and W Scotland), *Great Spotted Woodpecker, Wren, Robin, Mistle Thrush, Blackcap* (very rare N Scotland), *Chiffchaff* (very rare N and E Scotland), *Goldcrest, Long-tailed Tit, Carrion Crow.*

Commonest in South

Blackbird, Lesser Whitethroat (scarce Scotland, less common in Wales), *Blue Tit, Great Tit, Goldfinch* (absent from N and W Scotland), *Bullfinch.*

Commonest in East

Red-legged Partridge (very rare N and W of Yorkshire, Welsh border and Dorset), *Grey Partridge* (absent N and W Scotland, very rare Wales and SW England), *Pheasant* (absent N and W Scotland), *Moorhen* (very rare N and W Scotland), *Turtle Dove* (very rare Scotland, Wales, N and SW England, *Cuckoo, Skylark, Yellow Wagtail* (very rare Scotland, Wales, SW England), *Whitethroat* (absent N and W Scotland), *Yellowhammer, Reed Bunting, Corn Bunting* (almost absent in Scotland, Wales and SW England).

Commonest in West

Swallow, Garden Warbler (rare N Scotland), *Treecreeper, Jay* (absent N Scotland and S Uplands of Scotland), *Magpie* (almost absent from W and high ground in Scotland).

Commonest in North and West

Willow Warbler, Coal Tit, Chaffinch.

Commonest in North

Lapwing, Pied Wagtail, Spotted Flycatcher, Rook (absent N and W Scotland), *Starling.*

Common and widespread over all or most

Kestrel, Woodpigeon, Jackdaw (absent from N and W Scotland).

Table 4
Regional distribution (Winter)

Information is taken from Lack[2].

The six extra species:

Golden Plover (commonest in East), *Snipe* (commonest in South), *Black-headed Gull* (no particular concentrations), *Meadow Pipit* (commonest in South), *Fieldfare* and *Redwing* (both commonest in South and West).

The resident species:

Most species show similar patterns to the breeding season, although there is widespread movement away from high ground, and towards the south. The only species which show a major change are *Kestrel* (to mainly in East), *Lapwing* (to mainly in central England) and *Woodpigeon* (to mainly in East). The changes of the first two of these at least are probably due to immigrants from north and east[2].

References

1 O'Connor, R.J. & Shrubb, M. 1986. *Farming and Birds*. University Press, Cambridge.

2 Lack, P.C. 1986. *The Atlas of Wintering Birds in Britain and Ireland*. T. & A.D. Poyser, Calton.

3 Hardy, A.R., Stanley, P.I. & Greig-Smith, P.W. 1987. Birds as indicators of the intensity of use of agricultural pesticides in the UK. Pp. 119–132 in *The Value of Birds* (ed. A.W. Diamond & F.L. Filion). ICBP Technical Publication no. 6. International Council for Bird Preservation, Cambridge.

4 Marchant, J.H., Hudson, R., Carter, S.P. & Whittington, P.A. 1990. *Population Trends in British Breeding Birds*. British Trust for Ornithology, Tring.

5 Marchant, J.H. 1983. *BTO Common Birds Census Instructions*. British Trust for Ornithology, Tring.

6 Fuller, R.J., Marchant, J.H. & Morgan, R.A. 1985. How representative of agricultural practice are Common Birds Census farmland plots? *Bird Study* 32: 60–74.

7 Sharrock, J.T.R. 1976. *The Atlas of Breeding Birds in Britain and Ireland*. T. & A.D. Poyser, Calton.

8 Smith, K.W. 1983. The status and distribution of waders breeding on wet lowland grasslands in England and Wales. *Bird Study* 30: 177–192.

9 Shrubb, M. & Lack, P.C. 1991. The numbers and distribution of Lapwings *V. vanellus* nesting in England and Wales in 1987. *Bird Study* 38: 20–37.

10 Tucker, G.M. 1989. The winter farmland hedgerow survey. A preliminary report. *BTO News* no. 164: 14–15.

11 Smart, N. & Andrews, J.H. 1985. *Birds and Broadleaves Handbook*. Royal Society for the Protection of Birds, Sandy.

12 Green, R.E. 1988. Stone Curlew conservation. *RSPB Conserv. Rev.* 2: 30–33.

13 Potts, G.R. 1986. *The Partridge. Pesticides, Predation and Conservation*. Blackwell Scientific Publications, Oxford.

14 Greig-Smith, P.W. 1989. The Boxworth Project – environmental effects of cereal pesticides. *J. Roy. Agr. Soc. England* 150: 171–187.

15 Greig-Smith, P.W., Frampton, G.K. & Hardy, A.R. 1992. *The Boxworth Project: Pesticides, Cereal Farming and the Environment*. HMSO, London.

16 Shaw, P. 1988. Factors affecting the numbers of breeding birds and vascular plants on lowland farmland. *NCC Chief Scientist Directorate commissioned research report* no. 838. Nature Conservancy Council, Peterborough.

17 Bignal, E.M., Curtis, D.J. & Matthews, J.L. 1988. *Islay: Land types, bird habitats and nature conservation. Part 1: Land use and birds on Islay.* NCC Chief Scientist Directorate report no. 809, part 1.

18 Owen, M., Atkinson-Willes, G.L. & Salmon, D.G. 1986. *Wildfowl in Great Britain*. 2nd edition. University Press, Cambridge.

19 Matthews, R. (ed.) 1987. *Conservation Monitoring and Management*. A report on the monitoring and management of wildlife habitats on demonstration farms. Countryside Commission, Manchester.

20 Arnold, G.W. 1983. The influence of ditch and hedgerow structure, length of hedgrows, and area of woodland and garden on bird numbers in farmland. *J. appl. Ecol.* 20: 731–750.

21 Cooke, A.S., Bell, A.A. & Haas, M.B. 1982. *Predatory Birds, Pesticides and Pollution*. Institute of Terrestrial Ecology, Cambridge.

CHAPTER 2

Some General Principles

2.1   Range of habitats and patchiness

Farmland is a patchwork of different habitat features, although the largest area on a farm comprises the fields and other open ground. However, individual fields are of different sizes, and usually the boundaries are marked by such as fences or hedges, though treelines, stone walls, earth banks, ditches, rivers or tracks also occur. Hedges and fences are the most common and occur in nearly all regions, but some of the other types have pronounced regional concentrations. For example, walls occur particularly in areas where it was necessary to clear stones off the fields, e.g. the Cotswolds, and also on higher ground, and large water-filled ditches are in areas which are very low-lying, e.g. Somerset Levels.

Interspersed with the fields and their boundaries are several other habitat types, including woodland or coverts, scrub, ponds, lakes, rivers and streams and, of course, farm buildings. Some of these features are semi-natural, though usually modified, but others have been created by man, sometimes for a purpose which no longer applies. Rackham[1] provides a detailed history of the development of the countryside and the important features in it.

All this adds up to a great variety of habitat types and particular features, and there is therefore potentially a great variety of places for birds and other wildlife to occupy. Indeed habitat mosaics are beneficial to many birds, but there are some negative factors too. Firstly, most patches of habitat in farmland, except for open fields, are rather small, so that any animal or plant which requires a large area of a non field habitat is unlikely to find it there. The small size of patches does, however, mean that there is usually a good deal of 'edge' which tends to hold more birds and other animals than an equivalent area in the middle of a habitat type (see especially ch 8). A second difficulty is that habitat patches adjacent to fields are liable to be disturbed regularly, both physically by such as tractors, and from the drift or deliberate application of fertilizers or chemical sprays from fields. A third difficulty, but one which is perhaps less important for birds, is that, very often, patches of one habitat type are separated by areas of completely different habitat. Therefore some animals and plants may have difficulty in reaching the habitat patch even if it would be suitable when they got there. Birds are more mobile than most other groups of animals but some species, especially of woodland, seem to be reluctant to cross large open areas.

Individual farms vary in quality and availability of wildlife habitat. It is often true that a farm which has one interesting wildlife habitat feature will also have others of at least moderate interest. For example, a farm which retains large hedges will often have some rough untidy corners, and any ponds are probably less likely to have been drained. The reason is that all such habitat types depend on a degree of management sympathetic to conservation in order to retain that interest, and very often the level and intensity of the management over the farm as a whole is the critical factor.

It is worth emphasing again that it is usually easier, cheaper, and more effective, to manage and improve existing habitat features than to create them from anew. Even a small feature will have several animals and plants already present to form the basis for recolonisation. For example a pond which is partially filled in or a hollow where water collects will be a good site to renovate or build a pond as the soil is likely to be suitable for holding water and plants that prefer damp conditions may already be present. However, in areas where there is very little existing wildlife habitat any piece that is created will immediately become a haven for some individuals, although it is likely to take longer to develop and mature than an existing site. When trying to create a new feature, it is imperative also not to destroy some other valuable feature already present. For example an old wet rough field may be more valuable to conservation as a whole than any pond which could be dug to replace it.

In general, a variety of different habitat features and management methods will increase the numbers and diversity of birds on a farm. Details are spelt out in later chapters but there are benefits for birds of growing a range of crops in any one year, from leaving odd corners to mature naturally, managing the hedges in different ways and not all at once, encouraging a variety of native tree species to grow out of a hedge or renovating a wood, and leaving such as a pond to fill and mature naturally.

Over the centuries agriculture has changed considerably and has had its upturns and downturns, often dictated by economics, government incentives and policies, and the available means of carrying out various tasks. Since the Second World War there has been a major revolution in the methods of producing food. In particular, farm machinery has become much larger and more powerful, and there is now a variety of chemicals to deal with pests and other biological problems. The combination of these two factors in particular has meant that many natural restrictions on what can be grown and where have broken down. Therefore even the fields are now more uniform over a farm and over the country than they used to be.

Over the last few decades farms have become very much more intensive, and individual farms now tend to concentrate on one type of farming rather than being a mixed enterprise. There is also greater pressure to use all available land for production. This last has included the bringing in to production of what once were non-productive areas, and these were often the most useful for birds and other wildlife.

2.2 Regional differences

Within Britain there are pronounced regional patterns in farming practices and in the occurrence and structure of the different habitat features. As a result there are often different conservation priorities in different parts of the country. Detailed discussion is beyond the scope of this book but some information is necessary to help the reader interpret properly the chapters which follow.

The major sources of information about agriculture across Britain are the annual statistics published by MAFF and SOAFD. These give the area of each major crop grown, the numbers of livestock and several other factors relating to farming. They also give the area of other major land uses on agricultural holdings, e.g. woods and buildings, but they do not record other, often larger, areas of these away from such holdings.

The main contemporary difference in agriculture across Britain is that in the east the farming is predominantly arable, especially cereals, while in the west it is predominantly grass with cattle or sheep. Both occur in central areas but truly mixed enterprise farms are now much scarcer than they were even in the 1970s. Box 2.1 contains a map showing the proportions of tilled land and grass in 1987 and clearly shows the main east-west divide. There are also differences within regions which may affect some birds, for example the proportion of tilled land which is under cereals or other crops, and the proportion of grass which is rough grazing.

The pattern shown in this map also has implications for other aspects of farmland. For example where there are no livestock there is less agricultural justification for keeping hedges and other forms of shelter. There have always been fewer hedges in eastern England than elsewhere, and many have now been either severely trimmed or removed altogether. Low trimmed hedges are much less useful to birds than the larger ones which are predominant in western areas where they are used to keep in stock and provide shelter (see ch 3).

Box 2.1 Some Regional Patterns of Farming Practice in Britain

The first map shows counties (England and Wales) and regions (Scotland) which had a high proportion of their farmed area under tillage crops and those which had a high proportion under grass in 1987. The east-west divide is very obvious.

The second map shows the proportion of farm holdings which were classified in 1987 as woodland. Note that the figures are not the total area of woodland in a county but the amount on farm holdings.

Sources: MAFF and DAFS agricultural statistics for 1987.

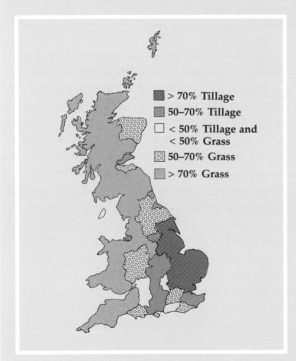

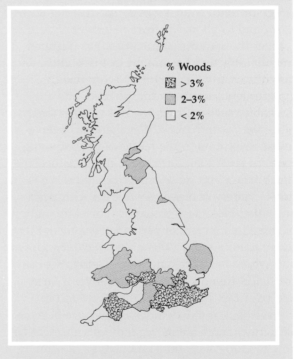

Box 2.1 also has a map showing the proportion of woodland on farm holdings. It is important to note that this is not the total area of woodland though it is related to it. It is clear that there is more farm woodland in southern England than farther north, which has implications for those birds that are most common in woods (see ch 8). Many of the woods are retained specifically as shooting coverts.

Other physical features also show some regional variation. Walls and large ditches are mentioned above (p. 8). Ponds have been filled in more often in those areas where there are no livestock needing the water to drink. In the eastern half of Britain buildings around farmhouses are often silos or grain stores, and usually inaccessible to birds, while in the west they are often animal houses and therefore more open and accessible. The plant species composition of hedges varies considerably. Indeed most habitat features show some variations across the country, whether due to climate, soil, topography or farming practices. Hence any bird species that are dependent upon particular features will vary regionally in their abundance resulting in geographical variation in the need for specific conservation measures.

Despite the regional variation in farming patterns many common farmland birds are widespread all over Britain, and often over large parts of western Europe as well. There are variations in density, some of which can be related directly to farming practices but many are related to other factors. Rather few bird species are actually restricted in their distribution to one part of the country. Box 1.1 gave some indications of where species are most common. Overall densities and number of species are higher in the south of Britain than the north with the southwest holding especially high figures for both. In general hedgerow and woodland species are more common in the southwest and field species in the east, but there are exceptions (Box 1.1 and ref 2).

In the following chapters some management measures are suggested which benefit particular bird species. Often, however, these measures will benefit a variety of other species as well. Where appropriate the suggestions are tailored specifically to particular areas but mostly they are widely applicable.

2.3 Economics of conservation in farmland

The initial impetus from farmers to find out about the possibilities for improving farmland and its associated habitat features for wildlife and conservation often comes from the desire to improve the shooting potential, or the aesthetic and amenity value of the farm. Luckily, management practices which improve these also often create good wildlife habitat and, although shooting is incompatible with some conservation objectives, both can usually be accommodated.

Farmers are often reluctant to commit money to enterprises when they cannot see the benefit in direct farming terms. However, many of the suggestions here do not cost any more than the farmer may have to spend anyway, being merely changes to the timing or extent of the operations. In addition, public demand is increasing and some of the management suggestions which are made in this book can attract grant aid for part or all of the work involved. Each scheme providing grants or loans is subject to particular restrictions and guarantees, and most have a limited overall life. They are difficult to generalise but practices for which money may be available include woodland management, shelter belt and hedge planting, pond renovation, and water resource management. Sources include MAFF, SOAFD, Forestry Authority, Local Authorities, and the Countryside Commission depending on the type of work envisaged. The addresses of these and other organisations are in Appendix 2.

References

1 Rackham, O. 1986. *The History of the Countryside.* J.M. Dent & Sons, London.

2 O'Connor, R.J. & Shrubb, M. 1986. *Farming and Birds.* University Press, Cambridge.

CHAPTER 3

Hedgerows

A 'hedge' comprises a field boundary that contains a line of shrubs or trees. In some parts of Britain there are stone walls and other structures at field boundaries but all these types are dealt with in the next chapter.

A typical hedge is a continuous line of bushes, but there is much variation across Britain in both structure and species composition. Many variations are caused by climate, geology and soil type, management (or lack of it), and the mixture of plant species, the last in turn related to the age of the hedge. There is no detailed information about how plant species composition varies in hedgerows in different parts of the country although there are distinctive types in a few places. For example, many hedges on Exmoor are of beech, many in the Breckland of East Anglia are of Scots pine, and many in Staffordshire and Herefordshire are of holly. Overall, the commonest shrub species in hedges is probably hawthorn. Sometimes this occurs on its own but there are often other shrubs mixed in, notably blackthorn, elder, field maple, beech, elm and hazel, often accompanied by ivy and bramble. For further details of structure, variations and history see for example the books by Pollard et al.[1], Dowdeswell[2] and Rackham[3].

Trees are a feature of many hedges and as with the shrubs, numbers and species composition vary across the country. The commonest are oak, ash, various willows and elm, although the last has become much scarcer since the destruction caused by Dutch elm disease in the 1970s. In a few places hedges are simply lines of trees without shrubs. These treelines are most prevalent in parts of East Anglia, where they have been planted as wind-breaks, with poplars or willows often being used.

Hedges probably hold a greater total number of breeding birds than any other feature in farmland. Thus they are arguably the single most important feature in determining the numbers of birds present. This may well be true of the winter too when the hedges are used especially as shelter by roosting birds. The actual bird density in a hedge is not easy to assess in a comparable way to the density in other features, because hedges are essentially narrow corridors and it is difficult therefore to define the area occupied. Nevertheless, it is clear that hedges are vital for bird conservation in farmland and any management which is carried out on them may have major effects on the birds in them. This chapter discusses the different forms of hedge, the different management techniques available and the effects of these options on the bird life.

Birds use hedges for many different purposes. Some species nest in them, some use them as elevated song perches, some roost in them, and many birds get some or all of their food from them. Invertebrates of many kinds and several species of berries are the most important foods. No bird species in Britain is confined to hedges, with the possible exception of the Cirl Bunting, which is now restricted to a few areas on the Devon and Cornwall coasts. (All but six of the 123 Cirl Bunting territories found in a survey in 1982 contained some hedgerow and the remaining six included some scrub[4]). Hedgerows, however, are a major breeding season habitat for several species in Britain, including Yellowhammer and Whitethroat[5], and are important at other seasons as well[6]. Many common and less common birds use hedges, and resident individuals may remain in the same area of hedge throughout their lives. The majority are species whose principal habitat is woodland, such as Blackbird, Chaffinch, Blue Tit, Song Thrush and Robin, others are mainly birds of scrub, such as Linnet, and a few are basically field species, such as Grey Partridge. All these birds, and others, have particular preferences for different structures and types of hedge, and it is clear that no single type of hedge is the best for them all, although there are some general features which are good for a wide variety.

Some types of hedgerow and some management practices are more beneficial to birds than others and there are excellent opportunities for making a positive contribution to the conservation of birds and other wildlife by looking after hedges. This is largely because it is easier to alter hedgerows than most other habitat features considered in this book. However, it is essential to think carefully about the full consequences and the objectives before embarking on any new management practices which will change the structure and composition of the hedge.

The rest of this chapter is divided into three main sections. The first considers the amount of hedgerow and its location in relation to other habitat features; it also includes an assessment of the effects of removing hedgerows. The second discusses the physical structure of a hedge and its plant species composition, i.e. what makes a 'good' or a 'bad' hedge from a bird's viewpoint. The third considers the various management practices which are used, and the benefits and drawbacks of each. At the end there are two very brief sections, firstly on the relationships with other features and secondly, specifically on planting new hedges. All three main sections are, however, closely interrelated and therefore should really be considered as a whole package. In particular, the aspects considered in the first two sections are usually the result of, or can be brought about by, applying the management practices of the third to different degrees and in different ways.

3.1 Hedges in the landscape

Amount of hedge per unit area

In the breeding season there are more birds and more species in hedges than in open fields. This may be true in winter too, although individual fields may contain large flocks of plovers, thrushes, Starlings and others (see chapter 6). Therefore, in general terms, farms with more hedges per unit area have more birds. However, there is no simple relationship and one cannot predict the actual overall density of birds simply from the amount of hedge present. For example, the geographical location of the farm (see ch 2) and the type of hedge (see below) are also very important. Box 3.1 illustrates schematically the relation between bird density and the amount of hedge.

The number of bird species present in an area also varies for many of the same reasons. As the amount of hedge increases so the size of the fields and open spaces decrease. This means that those birds which use open areas become less common. Hence the greatest number of species present in an area occurs with a lower amount of hedge than does the greatest density of individuals. Box 3.1 also illustrates this.

Each individual bird species has its particular optimum for the amount of hedge present. Most of the species which are primarily woodland birds, e.g. Robin and Blackbird, are more common where there is more hedge. In contrast, the field species show no very clear patterns, although most avoid the smallest fields, especially if these are bounded by high hedges or woods. Skylarks at least seem to avoid large areas of a uniform crop, preferring small areas of different crops[7]. Other species again are commonest at an intermediate hedge density. Little Owls seem to be commonest at a hedge density of about 40 m/ha[8] (equivalent to an average field size of 6–8 ha), and Grey Partridges at about double that[9]. In both cases this is a compromise between the need for cover for nest sites (holes for Little Owls, hedge

The accompanying graph gives a schematic representation of the overall density of birds and the overall number of species on farms with different amounts of hedge present. With more hedge the overall bird density rises, but the number of species peaks or levels off when fields become too small for open country species. The point where the number peaks or levels off will depend on the quality of the hedges and the amount and quality of other habitat features present. These in turn will depend on geographical location, soil type, exposure etc. The majority of farms will lie between the extremes illustrated.

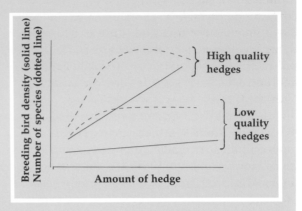

Source: the files of the Common Birds Census.

bottoms for the Grey Partridge) and open ground or crops as hunting or feeding sites.

In Switzerland the ideal for five species – Whitethroat, Blackcap, Garden Warbler, Red-backed Shrike and Yellowhammer – was found to be when hedges occupied about 4% of the total area, which meant fields no more than about 300 m across[10] (equivalent to about 60–80 m hedge/ha).

Similarly, in Finland the densities of twenty-seven bird species were compared between areas, with field sizes from 1–2 ha to 200 ha. The overall density of five species rose with increasing field size (Curlew, Skylark, Whinchat, Linnet and Ortolan Bunting), that of twelve species decreased (typical hedge species including Whitethroat, Magpie and Yellowhammer), and that of 10 species was fairly even through the range of field sizes or showed a variable relationship (mainly field species including Lapwing and Meadow Pipit, but also Sedge Warbler and Reed Bunting).

In winter many of the same factors seem to apply. Field species, such as Lapwing and Golden Plover, seem to prefer larger fields, i.e. fewer hedges, but for others which use both hedges and fields (e.g. thrushes) there is no clear relation[12].

Therefore, to retain a high density of birds and a broad range of species the density of hedges should not be allowed to fall below about 60–80 m/ha, which means an average field size of about 4–7 ha. Where there is more hedge than this there will be a higher overall density of birds but perhaps slightly fewer species, with those which prefer open areas dropping out. With fewer hedges, the overall bird density and number of species both fall.

The location of hedges with respect to other habitat features

Most hedges run between two fields or between a field and a road, but hedges also occur as the boundaries to woods, along so-called 'green lanes', and occasionally along rivers, streams and elsewhere. Green lanes, typically a hedge on either side of a track and which are often very ancient routes, hold a high density of birds (CBC data, personal observations). There is little or no information on any differences in numbers of birds along wood edges which contain hedges and those which do not because CBC maps do not usually distinguish between them. Hedges with a ditch alongside are preferred by some species, e.g. Song Thrush[13], and hedges on a raised bank are preferred by Grey Partridges[14], but associated fences or walls do not seem to add greatly to bird numbers in the hedge. See ch 4 for more details of these.

Hedges near to woods often contain more birds than those farther away. For example, on a farm CBC plot in Dorset, Chiffchaffs were recorded both along hedges and in small woods, the latter being the preferred habitat. When total numbers were low the majority of the records of birds in the hedges were in those close to the small copses. Hedges farther from the copses were only occupied when numbers were higher[15]. In Oxfordshire Long-tailed Tits were found to prefer to nest in hedges adjacent to woods[16]. This seemed to be because they were then nearer to good feeding areas in those woods, and yet suffered less nest predation than nests actually in the woods.

Breeding birds prefer sections of hedge near to intersections (where three or four hedges meet) than straight lengths of hedge[17]. In all there were 1.7 times more birds near intersections than along straight sections of the same total length. Box 3.2 illustrates the patterns for twelve common hedgerow species. Around intersections there is more hedge in a given area of ground. Therefore there may be more feeding opportunities, fewer costs entailed in defending a territory, and shelter from more directions than along straight sections. In my study the number of individual trees was the same near intersections as along straight sections[17], but in a different area there were 1.5 to 2 times as many plant species in a stretch near intersections[18].

The orientation of a hedge has implications for the temperature and degree of shelter it provides. A hedge running southwest to

northeast is likely to provide more shelter than others. In addition a farmer may be more likely to remove a hedge, or cut out the trees, if it runs east to west than one running north to south, because of increased shading on the north side.

In summary, hedges and sections of hedges adjacent to or near other habitat features are likely to hold more birds than those with simply a field on each side.

Hedges as links between other habitat features

It is often stated that some animals and plants, including birds, use hedges and other such 'lines' across open fields as corridors to get from one habitat feature to another. In Canada, small mammals and birds used fencerows as an entry into the fields to feed, though the birds did not seem to use them as a means of moving from one area to another[19]. Similarly in Devon, the only communication route for hedgerow birds between two areas separated by a 200 m playing field was along a line of elm trees[20]. The need for a corridor is perhaps more important for some other groups of animals than for birds, and some birds have been seen to occupy areas on both sides of a field as part of one territory, e.g. Robin across a 200 m field[21], and Blackbird even across rather more than that at times (CBC data). For at least some species a connecting hedge is obviously not essential.

Intuitively one would think that bird species which like woody vegetation and want to disperse would prefer to use a hedge line rather than cross an open field. Birds migrating over long distances are known to follow obvious features in a landscape such as rivers, lines of hills or the coast[22], and it seems quite possible that the same happens on a smaller scale within farmland. Casual observations suggest that, for example, Blue Tits and Great Tits move along hedges with their young after the breeding season, but it is not known how important the hedges might be.

Some circumstantial evidence for the use of hedges as corridors comes from a study of the

Box 3.2 Birds and hedge intersections

The average number of birds recorded along straight sections of hedge was compared with the average number along the same total length of hedge around intersections. On each of ten farm CBC plots records were counted along twelve straight sections of 75 m whose centres were at least 100 m from any intersection, and for 25 m along the three hedge stretches radiating from twelve intersections, each of which was at least 100 m from another. The diagram shows the average number of bird records along 75 m of hedge at each type over all ten farms for twelve common bird species. Clearly the numbers of most species are higher near to hedge intersections than along straight sections.

Source: Lack[17].

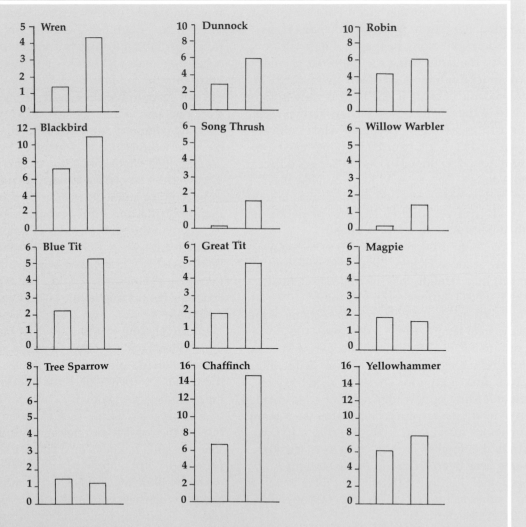

Average number of bird records along 75 m of straight hedge (left column) and around hedge intersections (right column) for 12 species

isolation of small woods on farmland in the Netherlands[23]. More bird species and individuals were found in woods surrounded by many hedges than in those which were otherwise similar, but were surrounded by few or no hedges – see ch 8 for more details.

In summary, there is very little definite information on the needs of birds for hedges as links and corridors. It may be much more important for less mobile animals and for plants. As such, farmers should if possible try to ensure that features of special wildlife interest, such as small woods, remain linked to each other by woody vegetation.

Hedgerow Loss

There has been a great deal of public comment and concern about the loss of hedgerows which has occurred in lowland Britain, especially since the end of the Second World War. There has also been argument about the effects on numbers of birds and whether or not any changes are significant. Rates of hedgerow loss over the country are discussed in Box 3.3.

Box 3.3 Hedge removal – when and where has it occurred?

Public concern about farmers removing hedges must be put into context. There were probably more hedges in the early years of the 20th century than there had been for several hundred years, so in one sense the removal was simply returning the landscape to a state which had existed previously. However, the last forty years have seen faster removal than ever before, although within this period the rate has not been constant over time nor evenly distributed across the country.

All the methods of assessing the rates of hedge removal have their biases[8], and there is a further difficulty because of arguments about the amount present at any one time (many recent studies use a baseline of 1945). However, by compiling all the available evidence, it seems that roughly a quarter to a third of the total length of hedgerow in the UK has been removed since then. It seems that the rate increased steadily through the 1950s, reached a peak of about 5500 km per year in the early 1960s and slowed somewhat from then through the 1970s[1,8]. Since 1980, several surveys have been carried out using a variety of methods, and over different

areas. The consensus seems to be that the rate of removal has certainly not slowed and may well be as high as it has ever been[24]. This last report, by the Royal Society for the Protection of Birds, also found that the average quality of hedges which are left is decreasing. A high proportion of them are low, narrow and intensively managed, which makes them much less useful as wildlife habitat (see sections 3.2 and 3.3).

The rate of destruction has not been constant over the country. There are still good agricultural reasons for keeping hedges in areas with livestock, as they provide shelter and prevent animals wandering. Around permanently arable fields, though, hedges have largely lost their agricultural use and chiefly have a landscape and amenity value, and may even be a net cost to the farmer. The result is that there has been a considerably greater rate of removal in arable areas than in livestock areas, which in turn means that eastern England has many fewer hedges than the western counties. In general the rate of hedgerow loss on arable farms has been 2–3 times that on mixed arable and pasture farms, and 4–10 times that on pasture farms[25]. However, on a geographical scale the greatest rates of loss, at least in the middle of the period being considered (the mid 1960s to mid 1970s), have been in the areas where there has been a progressive shift from mixed farming to a more intensive arable system. This picture is somewhat distorted by the fact that in eastern England there were many fewer hedges to start with.

The pattern is shown on the accompanying map. It indicates the average rates of hedgerow loss in England and Wales from 1964 to 1976. Areas on the

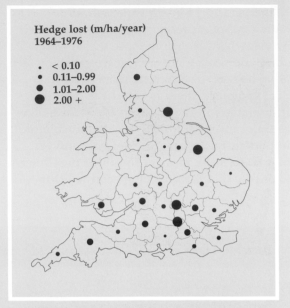

Hedge lost (m/ha/year) 1964–1976

- · < 0.10
- • 0.11–0.99
- ● 1.01–2.00
- ⬤ 2.00 +

western edge of the main cereal growing counties of eastern England were clearly subject to the greatest rates of loss, and it is probable that this is continuing. Data for the map are taken from Common Birds Census results, and each point gives the average of all data available for that county. The map is Figure 7.6 of *Farming and Birds*[8].

Surveys of hedgerow loss are also confounded by planting hedges. This has also occurred over the same period to varying extents both geographically and temporally.

Sources: mainly Pollard et al[1], O'Connor & Shrubb[8] and Joyce et al.[24]

The effects of hedge removal on birds depend on the amount of hedge present to start with and the amount and quality both of the hedges removed and those which are left. Results from several CBC plots which have been subjected to substantial hedge removal are presented in Box 3.4.

Summarising the results over several plots, it appears that hedge removal does not necessarily have major immediate effects. Numbers of individuals or species fell by no more than 20%, so long as some hedgerows or other suitable habitat features remained for the birds to live in. However, if the density of hedges falls below about 50 m/ha, bird numbers start to decline fairly quickly.

Another aspect of this is that, very often, hedgerow removal is accompanied by other agricultural changes, particularly intensification through a switch from mixed farming to predominantly or entirely arable fields. These other factors probably account

Box 3.4 The effects of hedge removal on birds

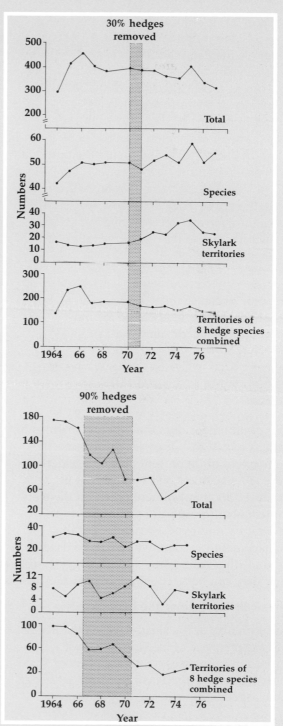

Farm 1 in Norfolk – information originally published by Bull *et al.*[26] but it has been re-analysed here in a slightly different way and allowance has been made for the national population trends over the relevant period.

In 1970–71 about 30% of the hedges were removed from the plot as part of an intensification to arable farming. The graphs show:

1 the total number of birds holding territory on the farm;

2 the total number of species holding territory;

3 the number of Skylark territories (as an example of a field species); and

4 the number of territories of eight resident hedgerow species combined (Wren, Dunnock, Robin, Blackbird, Song Thrush, Chaffinch, Linnet and Yellowhammer).

Numbers are shown for seven years on either side of the hedge removal although the data for 1969 are not included as it was a very atypical census due to rather few visits.

There was a slight decrease in the total numbers of birds recorded and in the numbers of hedgerow species, but a slight increase in the number of species holding territory and in the numbers of Skylarks.

Farm 2 in Cambridgeshire – information originally published by Evans[27], but as above the data have been re-analysed, and later years have been added.

From 1966–67 to 1970–71 the amount of hedge on the farm was progressively reduced from about 50 m/ha to about 3 m/ha. The graphs below plot the same figures as for Farm 1 over the period the farm was censused. In contrast to Farm 1 there was a considerable decrease in total numbers and in the numbers of hedgerow species, a slight decrease in the number of species overall and no change in the numbers of Skylarks.

Other Farms – Similar types of analysis on a further six CBC plots on which a third to two-thirds of the internal hedgerows were removed, showed that overall density fell by up to 20% on 5 plots and gained about 10% in the sixth, the number of species holding territory ranged from a gain of about 10% to a loss of about 15%, the number of Skylarks was relatively unaffected, and the numbers of the eight resident hedgerow species closely followed the overall total numbers.

Sources: Bull et al.[26], Evans[27], CBC data.

for some of the changes seen (see ch 6), but hedge removal is likely to be the most important factor for many bird species[25].

3.2   What is a good hedge for birds?

This section considers the different features of hedges which may increase or decrease the numbers of birds present. In most cases the different structures and features can be produced by management although there are constraints imposed by such as soil type, geography and exposure. Details of the management practices which can be used to produce them and some specific suggestions for how they can be adapted are in section 3.3.

Physical structure of the hedge

The structure of the shrub layer of the hedge and the presence of trees are the most important influences on the numbers of individuals and species of birds in that hedge. Each aspect is considered here in turn.

Size (height and width)
High hedges usually hold more breeding birds

than low ones, and wide hedges hold more than narrow ones. The actual denisty and number of species will, however, vary from one location to another depending on other local conditions. Some figures are given in Box 3.5.

The species most affected are those whose prime habitat is woodland. Blue Tit, Great Tit, Wren, Blackbird, Robin and Chaffinch are all much more common in higher and wider hedges (CBC data). Species whose primary habitat is scrub or very open woodland, such as Dunnock, Yellowhammer and Whitethroat, seem to be much less fussy about the structure of the hedges, and in one study from a CBC on a Dorset farm[33] the Yellowhammer and Whitethroat were found to be more common along low, trimmed, hedges than along well-timbered double hedges, with the Dunnock about equally numerous along the two types.

The field nesting species, typified by Skylark, prefer fields with few and low hedges. For example, Skylarks on a CBC plot in Hertfordshire only occurred in the largest field in years when the population was high. The reason appeared to be that this field was surrounded by high hedges with trees and there were also three large isolated trees

Box 3.5 Numbers of birds, numbers of species and breeding success in hedges of different sizes

A census was conducted on several parts of the Manydown Estate (Hampshire) in 1984[28]. The graph below plots the density of sixteen bird species (per km length of hedge) in tall hedges (usually at least 2 m high) against that in low hedges (mostly less than 1.5 m and often with gaps). On the graph the line indicates equal density along the two types; WR = Wren, D = Dunnock, R = Robin, B = Blackbird, M = Mistle Thrush, BC = Blackcap, WH = Whitethroat, WW = Willow Warbler, LT = Long-tailed Tit, CT = Coal Tit, BT = Blue Tit, GT = Great Tit, CH = Chaffinch, GR = Greenfinch, LI = Linnet and Y = Yellowhammer. Only three species were more common in low hedges and these only just so. The Chaffinch was found to be over three times as common in tall hedges, and the Blackbird nearly twice as common.

In hawthorn hedges in Cambridgeshire, those less than 1.2 m high supported fewer birds and fewer species than higher ones[29]. Similarly, Arnold[30] estimated that increasing the height of a hedge from 1.0 m to 1.4 m and the width from 0.8 m to 1.2 m, combined with not trimming the sides of the ditches every year and trimming only the year's growth off the shrubs, would lead to an increase of 30% in the number of breeding species and 400% in the total number of breeding territories in that hedge.

A study in SE Scotland[31] gave some figures for number of territories per km length for field boundaries (all of which contained a hedge) of different widths. The result is plotted below, redrawn from O'Connor[32]. As before, wider hedges held

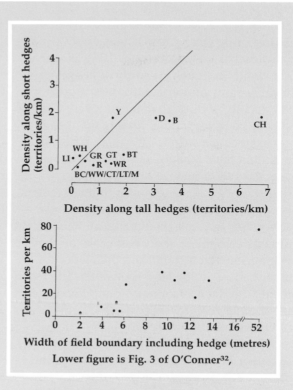

Lower figure is Fig. 3 of O'Conner[32],

many more birds although it must be noted that wider ones were also often higher.

Differences in breeding success in hedges of different sizes were not so clear. Parslow[29] found that breeding success was poorer in the lower hedges, probably due to rats. In contrast, Arnold[30] found that thrushes' nests were more successful in lower hedges, pointing out that nests in high hedges were more visible from above and the side. This last was, however, based on a rather small sample so may not be representative of all species.

Sources: Fuller[28], Parslow[29], Arnold[30], Shaw[31], O'Connor[32].

within it. Similarly, when Lapwings are roosting in the winter they also seem to prefer fields with low hedges[34,35].

Length of hedge

The length of a hedge, either between other features or between intersections, is not usually considered as an important variable. However, in lowland Scotland in a study of hedges of similar height, shorter lengths held a greater total density of birds than longer ones[36]. This is likely to be related to the finding that hedge intersections have up to 70% more birds near them than have straight sections (see above[17]). Inevitably, short hedges have a greater proportion of their length near to intersections. This preference was not consistent between species. For example, Dunnocks, all finches and buntings combined, and Chaffinches on their own, all showed a significant preference for longer hedges in lowland Scotland[36]. The result for the Chaffinch is contrary to my finding, and that of other researchers (e.g. ref 37), of a preference by Chaffinches for intersections. Perhaps there were other factors which made the longer hedges especially attractive to this species in the Scottish study.

Shape of the hedge profile

There has been a good deal of argument about the best shape for a hedge. It is clear from the above that larger hedges hold more birds than smaller ones, but whether the hedge is cut to an A-shape or a box-shape, or whether it is chamfered or rounded, appears not to be particularly important in itself for the majority of bird species. However, no detailed experiments have been done to prove this one way or the other. This aspect is considered again in section 3.3.

Livestock, especially cattle, often graze right into the edge of a field. As a result hedges around pastures may become tree-shaped, i.e. undercut and with very little herb layer vegetation. The livestock may also eat or damage lower branches of the hedge itself, particularly very close to the ground. This type of hedge tends to hold fewer birds[8], and nests

are also more liable to fail as, even in the upper parts, they are more visible from below to foxes and other ground predators. But again, critical figures are not available.

A similar 'undercut' hedge structure can accrue with the bad practice of spraying herbicides into hedge bottoms. The argument that such spraying is necessary to kill off weeds which will potentially invade the crop is rarely justified[38]. This is again discussed more fully under Management below.

Ground vegetation

The amount of ground vegetation under and immediately adjacent to the hedge is particularly important for some birds. Both species of partridge, the Pheasant and Reed Bunting all nest on the ground, and in farmland they prefer hedgerow bases. There are more nests of both partridges, and they survive better, where there is more ground vegetation present[39,14]. In addition several hedgerow species feed in this vegetation.

Gaps

Many hedges, especially narrower ones and those in arable areas where there is no necessity for them to be stock proof, have gaps in them. A gap may appear, for example, where a tree has died or been removed or in the immediate environs of a live tree because of shading. When the gaps only occupy a small proportion (up to about 10%) of the total length they are probably relatively unimportant to the numbers and diversity of birds present. However with more frequent and greater total length of gaps, fewer typical hedgerow birds are likely to occur simply because there is less woody vegetation available. Some species, such as the partridges and Skylarks, nest in thick grassy strips along field boundaries and they do not need any woody vegetation, but they are the minority. The next chapter (p. 37) has further discussion of such grassy strips as boundaries.

In summary, large hedges hold more birds though field species prefer lower hedges. Hedge shape seems relatively unimportant, except that ground vegetation is vital for some. Large gaps should be avoided.

Shrub species composition

The number of shrub species in a hedge is often used as a crude measure of the age of the hedge[1], with age (in hundreds of years) being approximately equivalent to the number of shrub species in a 30 yard (27 m) stretch. Each plant species has some particular characteristics and therefore a hedge with a mixture of species will be more diverse in the array of potential foods and niches available for birds and other animals. Some figures are in Box 3.6.

Considering individual shrub species, hawthorn hedges subject to a range of management methods consistently hold a higher density and more species of breeding birds than do elm hedges, with the largest numbers and the greatest difference in the least managed hedgerows[40]. There is little or no quantitative information on other shrub species, except that elder is generally regarded as holding rather few breeding birds, and that shrub species with a greater density of foliage

Box 3.6 The number of birds in hedges with different numbers of shrub species

The graph shows that the density of birds was higher where there were more shrub species in the hedge, although the variation found was considerable. (The graph is redrawn from Figure 7.2 of *Farming and Birds*[8].)

In Arnold's study[30] of some small East Anglian plots, the numbers of shrub and herb species present was one of the five most important habitat variables in determining the total numbers of birds present in both summer and winter, and individually of the numbers of Wrens, Dunnocks, and Robins in summer. In other words with more shrub species there were more birds.

A higher number of shrub species also contributed significantly to the increase in abundance of Wrens in a CBC study in Dorset[13].

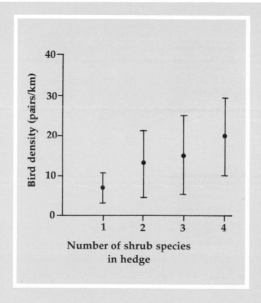

Sources: data from CBC files compiled by Leo Batten (see ref 8), Arnold[30].

Box 3.7 Birds and fruits

The table lists the major fruit-eating bird species with their favourite fruits in a study area in the Vale of Aylesbury

(Buckinghamshire), which comprised a mixture of farmland, scrub and woodland[41]. All bird species described as fruit-eaters are included, except those described as 'occasional or scarce'; and the Woodpigeon is included despite it being at least partly a seed-predator (see p. 171 of ref 41). All shrub species that accounted for at least 5% of the feeding records for that bird are listed, although not all occur in hedgerows. Each bird species clearly has its particular favourite fruits, with hawthorn and ivy perhaps the most important overall.

There are also some important seasonal variations. Some fruits are available all the year, but many more shrubs have ripe fruits between mid-July and January than in the remainder of the year. In addition, the time of depletion may be important for some birds. This depends on the volume of the crop, the preferences shown by the main fruit eating birds, and the times of arrival of some of the avian winter visitors. For example, when they first arrive in southern England in the autumn (exact timing is variable depending on conditions and available food further north), the Redwing and Fieldfare feed especially on haws in the hedges. Only later, from December onwards, when they have finished these do they switch to other berries, move out into the fields more, or move away altogether.

Elder is another very important fruit source. It appears in the lists for several bird species although it lasts only for a short while (in September). There are also a few specific associations not mentioned in the table, for example Bullfinches commonly eat ash seeds.

Fruit-eating birds and the fruits which occupied at least 5% of the feeding records. (The percentage of total records for that bird species is given in brackets.)

Woodpigeon ivy (67% (36% unripe, 20% ripe, 11% unspecified)), haw (14%), elder (8%).

Blackbird haw (23%), ivy (13%), holly (9%), rose (9%), rowan (7%), yew (7%), cherry (6%).

Song Thrush yew (26%), ivy (19%), blackthorn (16%), elder (9%), haw (6%).

Mistle Thrush holly (26%), mistletoe (17%), yew (16%), haw (10%), ivy (8%).

Fieldfare haw (59%), rose (26%), ivy (7%), holly (5%).

Redwing holly (32%), haw (29%), whitebeam (9%), ivy (9%), buckthorn (7%), dogwood (6%).

Robin spindle (21%), elder (16%), ivy (12%), dogwood (8%).

Blackcap ivy (23%), elder (20%), woody nightshade (8%), holly (8%), perfoliate honeysuckle (7%), spindle (7%), white bryony (6%).

Starling dogwood (43%), yew (18%), elder (16%), bramble (7%), ivy (6%).

Crows whitebeam (37%), elder (12%), wild cherry (12%), blackthorn (11%), dogwood (8%), crab apple (8%).

Other warblers elder (42%), perfoliate honeysuckle (14%), wayfaring tree (12%), bramble (8%).

Source: Snow & Snow[41].

in general hold more birds than those with less.

The other important effect of the shrub species composition on the numbers and species of birds present is that several shrubs produce fruits which are eaten regularly by birds and other animals during the autumn and winter. Box 3.7 lists the main fruit-eating bird species and their favourite fruits, and discusses some of the seasonal changes.

In general, more shrub species will mean more birds. The berries of some species, in particular hawthorn and ivy, are an important food source especially during the autumn and early winter.

Trees in hedges and treelines

There is a general increase in the numbers of birds where there are more trees in the hedge.

Certainly, the quickest way to reduce the numbers of birds in a hedge is to cut down any trees which are present. Conversely, planting trees or promoting tree growth by, for example, not cutting saplings when trimming the hedge is one of the best ways of increasing the numbers of birds.

An analysis of records of birds on sixteen CBC plots showed that there were 5.1 more records per 100 m along hedges with trees than along hedges without any on the same farms (a statistically significant difference).

Other studies have shown that trees are the most important feature of hedges for birds. In another analysis of CBC data, the numbers of more bird species were correlated with the amount of 'hedge with trees' than they were with any other habitat feature[8]. Similarly in Dorset[13], tree species diversity (using a figure combining the number of tree species and

---

Box 3.8 The numbers of birds along different types of hedges with trees

In Brittany the numbers of individual birds and species were determined along five different types of hedges with trees: (1) oak with dense shrubs underneath; (2) pollarded oaks; (3) oak with no shrubs; (4) gorse and broom; and (5) conifers. Clearly, those stretches with no shrub layer held many fewer birds than those with one.

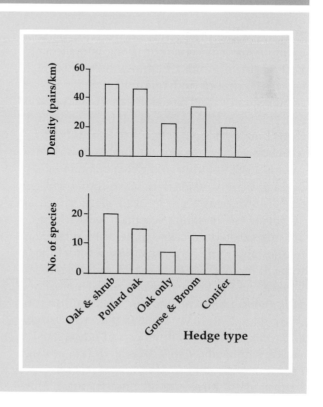

Source: Constant et al.[42], quoted by Baudry[43].

number of individuals) was the single most important habitat variable in determining the total number of breeding bird species and the total density of birds in the hedge. In this study, the numbers of some individual bird species were highly correlated as well, although other habitat variables were the most significant individual factors in some cases.

A line of trees on its own appears to be less attractive than one which has some shrubs underneath[8]. This is related to the point noted above, that hedges which have been undergrazed and where the shrubs have become tree-shaped hold fewer birds than managed thicker ones. Box 3.8 gives some figures for density and number of species along some different hedge types with trees in Brittany, and shows clearly that hedges with dense shrubs and trees hold many birds, and lines of trees without shrubs hold very few. The benefits of ground layer vegetation too are considerable for some birds (see above).

Species which are especially attracted to hedges with trees include Blackcap, Blackbird, Wren and Chaffinch (CBC data). Some individual species, however, seem to be indifferent and others seem actively to avoid them, being more common along hedgerows without trees. Among common hedgerow species, Dunnock and Yellowhammer are about equally common along those with and without trees (CBC data and ref 33), while the Corn Bunting seems actively to avoid them (CBC data). The majority of birds which live mainly in the fields avoid fields surrounded by high hedges, especially those containing trees (see above).

During the breeding season at least, birds are attracted specifically to the trees in the hedges and to those parts of hedges near the trees. Wren, Blackbird, Song Thrush and Chaffinch are recorded significantly more often within 25 m of a tree in a hedge than more than 25 m from a tree in the same hedge, and Willow Warbler, Blue Tit and Great Tit nearly significantly so (CBC data).

The reasons for this preference for trees are unclear. Most birds produce more young from

nests in shrubs than those in trees[44]. Trees provide extra vegetation structure and so potentially more feeding sites, but farmland birds are not often seen feeding in the trees. Treetops are often used as song posts and this may be the main attraction. However, the number of bird species occurring along a hedge in the winter is also correlated with the number of trees in that hedge[12], so there must be something else too. Perhaps the quality of the shrub vegetation is better in hedges with trees.

Distribution of trees along the hedge
The spacing of trees along hedges and any effects of this have not been studied. If a major use of the trees is as song posts, it is probable that trees at intervals or regularly spaced along the hedge (rather than all in one place) will provide the best distribution so far as increasing total numbers of birds is concerned. Also, a single tree is likely to be just as useful as two or more adjacent ones.

Size of tree
The effects of different sizes of tree are also unknown. It seems likely that larger trees will increase numbers of birds more than smaller ones, simply because larger trees create greater diversity. Yet even small trees, and saplings only a metre or so above the body of the hedge, are used regularly as song posts and may have the same effect as a larger tree.

Species of tree
There is much disagreement about the merits of different tree species. Native deciduous trees are generally thought to be preferred, although there is little real evidence. Willow (five species combined) and oak (two species) have higher numbers of insect species associated with them than any others in Britain and, in general, tree species which are indigenous hold more insect species than non-native ones[45]. Whether or not this means more individual insects, and therefore more potential bird food, is uncertain. Box 3.8 included a figure for coniferous hedges and showed that they support very few birds, although a few specialist species such as

Goldcrest and Coal Tit are more common in these than in broad-leaved equivalents. Conifer trees are not very common in hedges except for Scots pine in parts of East Anglia, although they do occur quite frequently in small woods or field corner plantings.

Several shrub species which produce berries, e.g. elder and hawthorn, do sometimes grow into trees. Other tree species listed in Box 3.8 also occur at times in hedges and their fruits attract birds when ripe. Fruits (or buds) of the two commonest trees in hedges, oak and ash, are only rarely eaten by hedgerow birds, however. Ash keys are a favourite food of Bullfinches and acorns are favoured by Jays, but neither of these birds is particularly common in hedgerows except when these fruits are available. The Jay is primarily a woodland bird and the Bullfinch prefers light scrub or woodland.

Dead trees

The benefit of trees in hedges is not confined to live trees. Dead ones are also important. This became a topic of considerable interest during the 1970s in the Dutch elm disease epidemic which killed a very large proportion of the elms over much of England. Osborne's analyses of CBC data and his specific work in Dorset was directed primarily at trying to find out the effects of this disease on birds[46,47]. He found that most bird species were not greatly affected by the death of the trees. Blackbird and Whitethroat decreased in numbers when the elms in the hedgerow died, but Goldfinch, Magpie and Stock Dove apparently favoured hedges with dead elms in them. However the subsequent felling of the dead trees led to the disappearance of eight bird species from the hedges and numbers of some others decreased. This in turn meant a general reduction in overall bird density and species diversity. The felling operation on Osborne's study farm also caused some damage to live trees and the opportunity was taken to do some other tidying up at the same time. All of these related factors added to the reductions in numbers of some birds[47].

One reason for the attraction for birds of dead trees is that, particularly if they have some bark remaining, a great many insects and other invertebrates will use them. Therefore they can become sought-after feeding sites, in particular by specialists such as the Great Spotted Woodpecker. Moreover, the death of a tree does not greatly alter the overall physical structure of the hedge, nor does it affect directly any creepers, in particular ivy. Many of the birds nesting on trees in hedgerows place their nests in ivy (Nest Record Scheme data).

In general, more birds and more species occur when there are more trees in a hedge, although a few birds avoid them. Trees spaced along the hedge may be more beneficial than all grouped together. It is likely that large trees are more beneficial than small ones, but small ones are seen to be used. The species of tree is probably less important than other factors. Dead trees can be important in themselves and for some purposes are probably as useful as live ones.

3.3 Management methods

The major methods of hedge management include mechanical trimming, which may be done annually or at intervals of several years, cutting back severely every five to ten years and perhaps with some light trimming in between, and traditional laying. This last method is normally done only every ten or more years, though often with some light trimming in between. The practical ways of doing these and other operations are fully described in a series of leaflets prepared by ADAS and others[48], and in a book published by the British Trust for Conservation Volunteers[49]. Some farmers hardly touch their hedges at all, leaving them to grow.

Other factors which influence birds are the timing of the operations, the frequency, whether or not all hedges on a farm are managed the same way, and whether all are treated at the same time.

Box 3.9 The numbers of birds along hedges subjected to different management regimes

The diagram (redrawn from Fig. 7.3 of *Farming and Birds*[8]) shows the density of birds (pairs/km length) along hawthorn (shaded bars) and elm (open bars) hedges subjected to different management practices ranging from little or no management at all to annual trimming. The number of species recorded along each type is noted at the bottom – those in brackets are from less than 900 m of hedge. Figures for rough grass and for post and wire fences are also included for comparison.

In a CBC study of a large plot, annually trimmed hedges held significantly fewer bird species than hedges managed any other way, regardless of the shape to which they were trimmed.[50].

Source: Table 9 of Moore *et al.*[40].

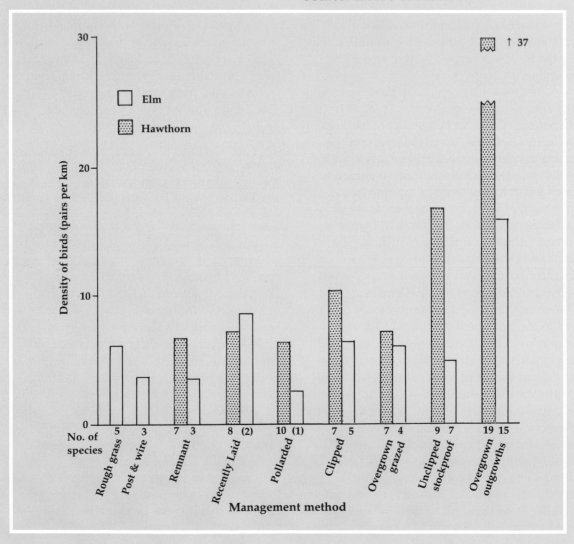

Some figures comparing the results of different hedge management methods are given in Box 3.9. Clearly, overgrown hedges with outgrowths from the main body of the hedge hold the largest numbers of individuals and species of birds by a considerable margin. However, unmanaged hedges do not necessarily retain high bird densities indefinitely. Hedges which are left completely untended over many years can become straggly and thin, especially near the base. This is particularly so in pasture areas where fields are grazed. In parts of Wales for example many individual hedges have become a line of scrubby trees often containing large gaps. These usually hold very few birds. Regular management is essential to retain a thick hedge but there is no information on the ideal regularity of it for birds, except for the partridge work mentioned below[14].

An interesting experiment is being carried out on a farm in Wiltshire, under the auspices of the Wiltshire Farming and Wildlife Advisory Group. On a 15 ha plot the hedges are being managed in a series of different ways. Each stretch of hedge is split into two contrasting management regimes and the effects on a variety of animals and plants, including birds are being monitored. The experiment is on rather too small a scale for the bird results to be easily interpreted, but for some other wildlife, notably invertebrates and plants, the results will be very useful to aid farm managers.

Methods of Management

Trimming

Trimming, usually carried out using a tractor-mounted cutter, is the commonest management practised, and if it is done every year to the same extent the breeding birds are unlikely to be affected. Exceptions are the two species of partridge, for which trimming every other year has been found to lead to more birds[14]. Under that regime more vegetation was left at ground level, but it was not allowed to become too rank to be suitable for nests. In order to retain some hedges which will be suitable every year under this regime, it is of course essential to trim the hedges on only half the farm each year.

The majority of hedges are subject to some trimming, and so long as it is done carefully it does not usually cause much damage to the hedge. There is the opportunity to adopt all kinds of shapes and size and the earlier sections of this chapter should be read with this in mind.

Periodic severe cutting

Some farmers leave their hedges untouched for some time and only manage them every 10 or more years. In such cases it is usually a drastic operation such as laying or what amounts to coppicing, although in most cases no wood crop is used by the farmer. Both of these operations temporarily change the structure of the hedge considerably, but in the longer term many hedges become thicker and therefore more suitable for most birds and wildlife if they are layed or 'coppiced' periodically. Some figures on the effects on birds of coppicing all the internal hedges on a Hertfordshire farm are given in Box 3.10. The short term effects were fairly catastrophic for the bird populations in the affected hedges, although most populations recovered during the subsequent five years or so[51].

Laying

Laying a hedge is likely to have similar effects on birds as coppicing, because they both result in a narrow hedge with few outgrowths, although a layed hedge usually remains stockproof. However, after a few years the layed hedge may well be thicker and be able to support more of the birds which need thick cover for nest sites, feeding or roosting. There has been, however, no specific comparison of the short or long term effects of laid and unlaid hedges.

Spraying

When fertilizers or pesticides are used on a field it is important to avoid any drifting of the chemicals into the hedge. Fertilizers do not affect birds very much directly but they may have considerable effects on the plants. In

Box 3.10 The effects on birds of 'coppicing' hedges

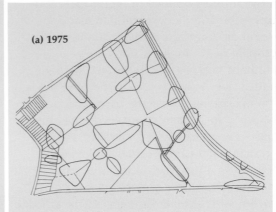

(a) 1975

In the winter of 1975/76 all the internal hedges on a CBC plot in Hertfordshire were cut back severely. For several years they had been largely neglected and were 3–4 m high with many outgrowths. After the cutting most of the hedges were essentially a series of more or less isolated stumps about 0.5 m high, except for three stretches containing lines of trees which were left untouched. Subsequently the hedges were again left unmanaged, and by 1985 they were in more or less the same state as in 1975.

The graph shows the number of bird

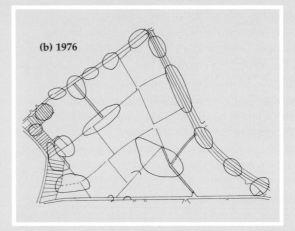

(b) 1976

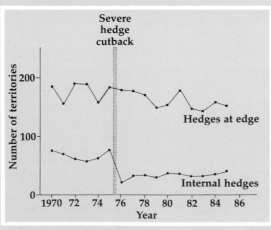

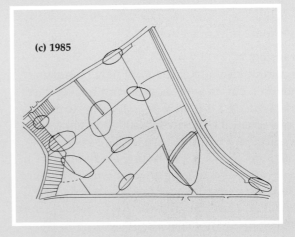

(c) 1985

territories recorded on the farm over several years along the edge hedges and along the internal boundaries. The graph is redrawn from Figure 2 of Lack[51]. Clearly, numbers along the internal hedgerows were considerably reduced after the cutting.

Numbers of some individual species were simply reduced, but for others the immediate effects were somewhat alleviated by some of the birds being able to relocate into surrounding uncut hedges, although there is no information on whether they achieved an equivalent nesting success or survival at this higher density. The maps of the distribution of Dunnock territories shown here are an example. In 1975 there were 9 or 10 territories along the internal hedgerows but this was reduced to only two in 1976, and both of these were partially associated with tree lines (indicated on the maps by the thicker boundaries). Many more were located in the boundary hedgerows. By 1985 the distribution had returned more or less to the 1975 pre-coppice arrangement, although the total number of territories had declined (eighteen to ten) in line with a decrease in the national population level.

Source: Lack[51].

particular, they may reduce the number of plant species that are present by promoting the growth of competitive species which then swamp the less competitive ones. Pesticides are different. If used incorrectly they kill animals, plants or both, and thereby remove either some of the shelter or some of the available food.

The practice of deliberately spraying herbicides into the base of hedges is also to be avoided. Not only does this reduce the amount of vegetation available for use as nest cover by several species, but it also reduces the overall shelter provided by the hedge which will affect other species too. Such spraying is often done on the pretext of removing weed plants which might invade the crop, and 60% of arable farmers do do it[38]. However, very few noxious weeds occur widely in hedges[38], and the practice is therefore usually unnecessary. Specific spot treatment may be required for such as cleavers or thistles, but this is unlikely to do any widespread harm to wildlife.

Deliberate spraying of herbicides into hedge bases can very often actually be counter productive[38]. Spraying out the plants creates more open ground which is ideal for annual weeds. These are usually the weed species which cause the problems in crops whereas perennial species are usually less problem.

Perennials are favoured by leaving vegetation unsprayed.

In summary, some management is necessary to retain the numbers of birds in hedges. Trimming done every year will have little or no effect on bird numbers. Laying or 'coppicing' have severe short term detrimental effects, but in the longer term are beneficial. Spraying of hedge bases with herbicides is usually unnecessary and should be avoided.

Timing and frequency of management

A critical factor for birds in respect of trimming and other management is the timing of the operation. For example, on hawthorn and blackthorn the berries are usually most abundant on the outermost twigs, and particularly on outgrowths from the main body of the hedge. It is these twigs which are cut off by the trimmer, and trimming will therefore deprive birds and other animals of an important food supply if it is carried out before the berries have been eaten. On arable land farmers now try to do as many operations as possible in the autumn period after harvest, before sowing the next crop and the land becoming too wet to drive over easily. The practice has increased with the modern prevalence of autumn sowing. However, the ideal time from the birds' viewpoint for hedge trimming and other hedgerow management work is January or February, after the berries have been eaten but before the birds have started to nest and the plants have started to grow again. It has also recently been found that fruits are much less common below 2 m from the ground and on bushes which have been cut back at all in the previous twelve months[52]. So for this reason too hedges should not be trimmed every year.

It is clear from earlier sections that a range of hedge structures present within an area at one time is likely to be the most beneficial for birds in general. Therefore the best regime to operate is to work on only a proportion of a farm's hedges in any one year. This is especially the case with the most severe operations as many of the displaced birds will be able to remain, albeit at a higher density, in those hedges which are not cut then.

In summary trimming is best carried out in January or February after the berries have been eaten. To retain a high bird diversity, only a proportion of the hedges on a farm should be worked on in any one year.

3.4 Hedges in combination with other features

Many hedges have a ditch running alongside which may or may not contain running water, some have a fence incorporated into the structure, some have a strip of grass or other vegetation alongside between the hedge and the crop, and some are associated with walls and/ or banks. All these other features, both on their own and in combination, are considered in the next chapter.

3.5 Planting a new hedge

All the above sections should be considered when planting new hedges, but there are a few extras as well:

a if there is a choice of site it is likely to be of most benefit to birds if it connects one or more existing habitat features;

b the best shrubs to plant are likely to be those which are already growing well in other local hedges, with the best specific mixture depending on land type, exposure and location;

c to get a thick hedge in a short time, it is normally best to plant two rows of shrubs;

d the hedge will hold more birds later on if some trees are planted at the same time and allowed to grow into standards; and

e some careful use of fertilizer may enhance growth, especially in the early stages.

Before they become fully established, young shrubs may need some protection especially from grazing animals. Farm livestock can usually be kept away by fences, either permanent or temporary, but rabbits may necessitate more substantial measures. Netting around the whole area is one solution, and protective guards around individual plants another. It is also important, especially with younger plants, to ensure that any herbicide sprays are kept well away, although some careful removal of herbaceous weeds from around the base of shrubs can help their growth. This may be done with selected and selective herbicides, although physical weeding or placing a black plastic sheet around the base will have a similar result. Further details and more specific advice on techniques and plant species are available from ADAS[53], the British Trust for Conservation Volunteers[49] and others.

3.6 Summary and Management Recommendations

General

1 Overall, hedges probably hold more birds than any other habitat feature on farmland, although the density is difficult to measure in a comparable way because of the linear nature of the hedge. Birds use hedges for many purposes – nests, roosts, shelter, and as food sources, in particular for invertebrates and berries.

Amount and location of hedges (Section 3.1)

2 The ideal amount of hedge on the ground is different for individual bird species. In general, the total density of birds in an area is higher when there are more hedges, but the number of species seems to be highest at a density of about 70–100 m/ha. This is equivalent to an average field size of about 5 ha. Numbers of birds are considerably reduced when the hedge density falls below about 50 m/ha and somewhat reduced below 80 m/ha. Field species prefer areas with fewer hedges.

3 Stretches of hedge near intersections hold more birds than straight stretches. Stretches near to woods also hold higher numbers.

4 Hedges linking other habitat features are the most useful in the landscape as a whole, because of their use as corridors.

Hedge structure (Section 3.2)

Physical

5 Tall, wide hedges hold more breeding birds than short, narrow ones, though this does not apply to all bird species individually. Field species in particular prefer areas with low hedges. Hedgerows are much less useful to birds if they are less than about 1.5 m high and 1 m wide. Shorter lengths with several intersections hold greater numbers than single long stretches although some species prefer the latter.

6 The shape of the profile seems relatively unimportant to most birds. However, hedges which are undercut, with the shrubs becoming tree shaped and with little vegetation near the ground, hold many fewer birds.

7 Ground vegetation is very important for several species, for example as safe nest sites for the two partridges. Some grass and herbs should be retained under and adjacent to the hedge, and the area should not be sprayed with herbicides.

8 Gaps should not be allowed to become too large as they reduce the shelter provided by the hedge to both birds and livestock.

Shrub species

9 Hedges with a greater variety of shrub species hold more birds. Hawthorn is the commonest individual species in many areas and holds more birds than elm. In general those shrubs with dense foliage hold more. Several shrub species have fruits which are relished by birds, especially in autumn. Hawthorn, blackthorn, elder, ivy and holly are amongst the favourites of several bird species.

Trees in hedges

10 Encouragement of trees in hedges is one of the best ways of increasing numbers of birds. The easiest method is to leave saplings to grow up through.

11 Trees spaced along a hedge are likely to be more beneficial than if all are clumped in one place.

12 Even small trees only a few metres above the body of the hedge can make a substantial difference.

13 Indigenous broad-leaved trees hold more insects and therefore potentially attract more birds than non-native broad-leaved trees. Conifers hold rather few, except for a few specialists.

14 Dead trees have many of the benefits of live ones. They also attract some birds, e.g. woodpeckers, which use dead bark as feeding sites.

Management (Section 3.3)

15 To retain a high diversity of birds, some management is usually necessary. Hedges which are left to grow unchecked, especially those which become straggly and undercut, are much less useful to birds.

Methods and timing

16 Regular trimming will have little direct effect on breeding birds. For partridges at least, trimming only in alternate years provides the best nest sites, and fruits are more abundant if the bushes have not been trimmed in the previous 12 months. Late winter and early spring are the best times for trimming to be carried out as most of any berries present will have been eaten.

17 Managing on a longer cycle of severe cutting (or laying) followed by several years of little, if any, management will have severe detrimental short term effects but beneficial longer term effects on bird numbers. To alleviate some of the short term ones it is best to cut or lay only some of the hedges in any one year. Many birds will live temporarily at higher densities in unmanaged hedgerows.

18 Spraying into hedge bottoms with herbicides is usually unnecessary and should be avoided.

19 When managing the hedges in an area the ideal is for there to be a variety of structure present at any time. This is often best achieved by managing on a rotation rather than by working on all hedges every year.

Planting new hedges (Section 3.5)

20 Many of the points made have implications for planting new hedges. These may in addition require some guarding against grazing by livestock or rabbits, and may require some careful fertilizing and removal of weeds from their surrounds.

References

1 Pollard, E., Hooper, M.D. & Moore, N.W. 1974. *Hedges*. Collins, London.

2 Dowdeswell, W.H. 1987. *Hedgerows and Verges*. Allen & Unwin, London.

3 Rackham, O. 1986. *The History of the Countryside*. J.M. Dent & Sons, London.

4 Sitters, H.P. 1986. Cirl Buntings in Britain in 1982. *Bird Study* 32: 1–10.

5 Williamson, K. 1967. The bird community of farmland. *Bird Study* 14: 210–226.

6 Osborne, P.E. & Osborne, L. 1985. A winter bird census on farmland. Pp. 287–299 in *Bird Census and Atlas Studies* (eds K. Taylor, R.J. Fuller & P.C. Lack). *Proc. VIII Int. Conf. Bird Census and Atlas Work*. British Trust for Ornithology, Tring.

7 Schläpfer, A. 1988. Populationsökologie der Feldlerche *Alauda arvensis* in der intensiv genutzen Agrarlandschaft. *Orn. Beob.* 85: 309–371.

8 O'Connor, R.J. & Shrubb, M. 1986. *Farming and Birds*. University Press, Cambridge.

9 Potts, G.R. 1980. The effects of modern agriculture, nest predation and game management on the population ecology of patridges (*Perdix perdix* and *Alectoris rufa*). *Adv. Ecol. Res.* 1: 1–82.

10 Pfister, H.P., Naef, B & Blum, H. 1986. Qualitative und quantitative Beziehungen zwischen Heckenvorkommen im Kanton Thurgau und ausgewählten Heckenbrütern: Neuntöter, Goldammer, Dorngrasmücke, Mönchgrasmücke und Gartengrasmücke. *Orn. Beob.* 83: 7–34.

11 Piironen, J., Tiainen, J., Pakkala, T. & Ylimaunu, J. 1985. Suomen peltolinnut 1984. [Birds of Finnish farmland in 1984]. *Lintumies* 20: 126–138.

12 Tucker, G.M. 1989. The winter farmland hedgerow survey. A preliminary report. *BTO News* no. 164: 14–15. And personal communication.

13 Osborne, P.E. 1984. Bird numbers and habitat characteristics in farmland hedgerows. *J. appl. Ecol.* 21: 63–82.

14 Rands, M.R.W. 1987. Hedgerow management for the conservation of partridges *Perdix perdix* and *Alectoris rufa*. *Biol. Conserv.* 40: 127–139.

15 Osborne, P.E. 1982. *The effects of Dutch Elm Disease on farmland bird populations*. Unpublished D. Phil. thesis, Oxford University.

16 Gaston, A.J. 1973. The ecology and behaviour of the Long-tailed Tit. *Ibis* 115: 330–351.

17 Lack, P.C. 1988. Hedge intersections and breeding bird distribution in farmland. *Bird Study* 35: 133–136.

18 G. Fry personal communication.

19 Wegner, J.F. & Merriam, G. 1979. Movements by birds and small mammals between a wood and adjoining farmland habitats. *J. appl. Ecol.* 16: 349–357.

20 Blackwell, J.A. & Dowdeswell, W.H. 1951. Local movement in the Blue Tit. *Brit. Birds* 44: 397–403.

21 M.R. Fletcher personal communication.

22 Mead, C. 1983. *Bird Migration*. Country Life/ Newnes Books, London.

23 Van Dorp, D. & Opdam, P. 1987. Effects of patch size, isolation and regional abundance in forest bird communities. *Landscape Ecol.* 1: 59–73.

24 Joyce, B., Williams, G. & Woods, A. 1988. Hedgerows: still a cause for concern. *RSPB Conserv. Rev.* 2: 34–37.

25 Hooper, M.D. 1977. Hedgerows and small woodlands. Pp. 45–57 in *Conservation and Agriculture* (eds J. Davidson & R. Lloyd). John Wiley & Sons, Chichester.

26 Bull, A.L., Mead, C.J. & Williamson, K. 1976. Bird-life on a Norfolk farm in relation to agricultural changes. *Bird Study* 23: 163–182.

27 Evans, P. 1972. The Common Birds Census: eight years at Ely. *Cambridge Bird Club Report* 45: 36–39.

28 Fuller, R.J. 1984. *The distribution and feeding behaviour of breeding songbirds on cereal farmland at Manydown Farm, Hampshire, in 1984.* Report to the Game Conservancy. Research Report no. 16. British Trust for Ornithology, Tring.

29 Parslow, J.L.F. 1969. Breeding birds of hedges. *Monks Wood Experimental Station Report* 1966–68: 21.

30 Arnold, G.W. 1983. The influence of ditch and hedgerow structure, length of hedgerows and area of woodland and garden on bird numbers in farmland. *J. appl. Ecol.* 20: 731–750.

31 da Prato, S.R.D. 1985. The breeding birds of agricultural land in south-east Scotland. *Scott. Birds* 13: 203–216.

32 O'Connor, R.J. 1987. Environmental interests of field margins for birds. Pp. 35–48 in *Field Margins* (eds J.M. Way & P.W. Greig-Smith). British Crop Protection Council Monograph no. 35, Thornton Heath.

33 Williamson, K. 1971. A bird census study of a Dorset dairy farm. *Bird Study* 18: 80–96.

34 Shrubb, M. 1988. The influence of crop rotations and field-size on a wintering Lapwing *V. vanellus* population in an area of mixed farmland in West Sussex. *Bird Study* 35: 123–131.

35 Barnard, C.J. & Thompson, D.B.A. 1985. *Gulls and Plovers: the Ecology and Behaviour of Mixed-species Feeding Groups*. Croom Helm, London.

36 Shaw, P. 1988. *Factors affecting the numbers of breeding birds and vascular plants on lowland farmland.* NCC Chief Scientist Directorate commissioned research report no. 838. Nature Conservancy Council, Peterborough.

37 Newton, I. 1972. *Finches*. Collins, London.

38 Marshall, E.J.P. & Smith, B.D. 1987. Field margin flora and fauna; interaction with agriculture. Pp. 23–33 in *Field Margins* (eds J.M. Way & P.W. Greig-Smith). British Crop Protection Council Monograph no. 35, Thornton Heath.

39 Potts, G.R. 1986. *The Partridge. Pesticides, Predation and Conservation*. Blackwells Scientific Publications, Oxford.

40 Moore, N.W., Hooper, M.D. & Davis, B.N.K. 1967. Hedges I. Introduction and reconnaissance studies. *J. appl. Ecol.* 4: 210–220.

41 Snow, B. & Snow, D. 1988. *Birds and Berries*. T. & A.D. Poyser, Calton.

42 Constant, P., Eybert, M.C. & Maheo, R. 1976. Avifaune reproductrice du bocage de l'Ouest. Pp. 327–332 in *Les Bocages: Histoire, Ecologie, Economie*. INRA, CNRS, ENSA and Universite de Rennes.

43 Baudry, J. 1988. Hedgerows and hedgerow networks as wildlife habitat in agricultural landscapes. Pp. 111–124 in *Environmental Management in Agriculture* (ed. J.R. Park). Belhaven Press, London.

44 Lack, P.C. 1988. Nesting success of birds in trees and shrubs in farmland hedges. *Ecol. Bull.* 39: 191–193.

45 Kennedy, C.E.J. & Southwood, T.R.E. 1984. The number of species of insects associated with British trees: a re-analysis. *J. Anim. Ecol.* 53: 455–478.

46 Osborne, P.E. 1982. Some effects of Dutch Elm Disease on nesting farmland birds. *Bird Study* 30: 27–38.

47 Osborne, P.E. 1985. Some effects of Dutch Elm Disease on the birds of a Dorset dairy farm. *J. appl. Ecol.* 22: 681–692.

48 ADAS. 1982. *Managing farm hedges*. Leaflet 762. Ministry of Agriculture, Fisheries and Food, London.

49 British Trust for Conservation Volunteers. 1975. *Hedges—a practical conservation handbook*. BTCV, Wallingford.

50 Ward, E.M. 1989. *An investigation of the relationship between farming practices and selected farmland bird populations using Common Birds Census data*. Unpublished M.Sc. thesis, Imperial College, London.

51 Lack, P.C. 1987. The effects of severe hedge cutting on a breeding bird population. *Bird Study* 34: 139–146.

52 Moorhouse, A. 1990. Managing farm hedges. *Fourth Report of TERF (The Environmental Research Fund)*: 14–16.

53 ADAS. 1986. *Hedgerows*. Advisory Leaflet no. P3027. Ministry of Agriculture, Fisheries and Food, London.

CHAPTER 4

Other Field Boundaries

Field boundaries are one of the most important habitat features for birds and other wildlife on a farm. Hedges are the most common, the most variable and the most studied type, and have therefore warranted a chapter to themselves (ch 3). However, other kinds, either on their own or in combination with a hedge or each other, may be the dominant type in some areas and may be important for birds. All types harbour a few and they will be affected by any management which is carried out. In this chapter, rough grass strips, fences, walls, banks and ditches are discussed in turn.

4.1   Rough grass strips

Most field boundaries contain some grass and this may be an important part of the boundary (see ch 3 and below). In some places, especially in parts of southern and eastern England where there is no need for a stock-proof barrier, a grass strip may be the only boundary between adjacent arable fields.

The only bird species which regularly place their nests in such grass strips are the Grey Partridge and Skylark, although the Red-legged Partridge and a few others will do so at times and especially if there is some woody vegetation nearby. Some specific information is given in Box 4.1. Very simply, these species prefer a strip which is wide enough (normally at least a metre), and with the standing grass and other plants of sufficient height (25–35 cm at least) to prevent incubating birds from being seen easily by passing predators such as foxes.

Grass strips are also used by other birds as foraging areas. In otherwise intensively arable areas of eastern England, strips which are part or all of a field boundary or those which are part of a headland are considered to be

important to Barn Owls[6]. For the owls the strips need to be at least 5 m wide, and those adjacent to a watercourse or a ditch are thought to be the most useful of all, because they harbour more voles and other small mammals which are the main food of the species. Those strips with a hedge or fence alongside are also preferred because they offer the owls the alternative of hunting from an elevated perch as well as quartering the ground from the air. For an area to support Barn Owls on a permanent basis there must also be a suitable nest site within a few hundred metres (see ch 11).

The Barn Owl has been receiving special attention from conservation organisations because it has been declining rapidly over the last 50 years. The reasons are uncertain but the loss of potential feeding sites, such as grass strips bordering fields, may well be more important than loss of potential nest sites[6].

All the species mentioned so far prefer the grass to be at least 25 cm or so high. However, many of the species which habitually feed on grassland, such as thrushes, Starlings, crows and (in winter) plovers, prefer grass which is either grazed or mown and therefore shorter than this (see ch 6). Grazing is obviously impossible in arable areas but mowing could be done, although most of such species are unlikely to feed on narrow grass strips anyway, unless they are fairly close to hedges or other woody vegetation. A direct experiment on the use of mown and unmown roadside verges by different bird groups was carried out in Denmark. There was no difference in the use by Skylarks or any other group[3].

Clearly, there is no advantage to birds in mowing such a strip. However, boundaries

Box 4.1 Grass strips as nesting habitat for partridges and Skylarks

Both species of partridge and the Skylark place their nests on the ground in a well-concealed position, and all rely mainly on camouflage to avoid detection. In addition, they all sit very tight in the presence of predators and only vacate a nest at the last moment, or often too late in the case of the Grey Partridge. On one site about a quarter of all incubating female Grey Partridges were taken by foxes, and this was despite an organised campaign against them[1]. Obviously, a bird nesting in a strip which is very narrow will be more visible to a predator passing along beside it than will a nesting bird in a wider strip.

Nest sites selected by Grey Partridges contained more dead grass, more bramble and more leaf litter than expected from the availability of these, and any earth bank present was higher. Sites selected by Red-legged Partridges had more dead grass, bramble and leaf litter, and also more nettles[2]. Any woody vegetation was less important than these features for either species. The largest amount of dead grass and other vegetation needed for ideal nest sites was most abundant in those boundaries which were cut biennially. So in order to maintain the highest density of partridges on a farm only half the boundaries should be managed (mowed or trimmed etc) each year[2].

In a study area in Denmark which contained very little woody vegetation, Skylarks were found to nest on roadside verges four times more often than expected from their availability, and to use them as foraging sites twice as much as expected[3]. The roadside verges in this case were grass strips, often including a ditch, but without hedges. The reason for the preference was partly that the fields on this site contained mainly spring-sown cereals, so there was little or no green cover when the Skylarks began setting up their territories. But it was evidently not the whole story as similar figures were obtained in East Anglia, where Skylarks were more often found on 5 ha study sites if there was a grass verge or a ditch than if there was not[4], and in Switzerland where the species was found more commonly on farms with small plots of different crops than on those with large areas of monoculture[5].

Sources: mainly Potts[1], Rands[2] and Laursen[3].

which were left unmanaged over several years held significantly less grass and other ground vegetation than all other types except grazed edges[2], while those cut in alternate years held the most.

Grass strips across fields are currently the subject of research by the Game Conservancy Trust. The strips hold much higher overwintering populations than fields of insects which are potential predators of pests such as aphids[7]. All these insects could provide food for birds as could the seeds of grasses or other weeds which might grow in the strips. Clearly, these benefits will disappear if pesticides are sprayed into the strips, just as spraying into hedge bottoms has detrimental effects (see ch 3, p. 31).

Birds which use grass strips prefer some growth to provide sufficient cover for them or their prey. The largest amount of ground vegetation is in hedges and boundaries managed in alternate years. Strips should be wide enough for nesting birds to be hidden from predators passing along the edge.

4.2 Fences

Fences are hardly used by birds – see Box 4.2. This is perhaps not surprising because fences provide neither shelter nor (directly) any food. In fact, as is clear from the CBC farm described in the Box, the only uses are as song posts or as vantage points to look for prey, depending on the species.

If the fence is a long way from any woody vegetation the only bird likely to use it as a song post is the Corn Bunting, a species which is now fairly local and largely limited in its range to the eastern half of Britain (ref 8 and Box 1.1).

The commonest species to use a fence as a vantage point is the Carrion Crow, which will only be present if there is likely to be potential prey such as small songbirds, small mammals or large invertebrates available in the field margin or the crop. Magpies and occasionally Kestrels, Little Owls or Barn Owls will also use them in this way. Therefore putting up a fence on a wide grass strip or similar field boundary, especially where partridges or Skylarks might nest, will overall probably harm birds more than benefit them.

Fences within a hedge have negligible, if any, direct effect on birds.

On balance therefore fences will be detrimental to birds because they are used as look-out posts by potential predators. However they are sometimes used as songposts by other species.

4.3 Stone Walls

In many parts of the uplands and in some areas of lower ground (e.g. south-west England) field boundaries are often marked by stone walls, especially drystone walls. They provide considerably greater shelter, to both stock and wildlife, than most other types of field boundary. Despite this, no birds seem to be dependent on them, although a few do use them within the regions of Britain where stone walls are the characteristic field boundary type. An example from a CBC plot in Derbyshire is cited in Box 4.3.

The Wheatear is perhaps the species which most commonly nests in stone walls. It needs a suitable crevice, with the nest itself normally out of direct sight of would-be predators. But many walls, especially those which are well made and maintained, do not have a suitable site. Pied Wagtails behave similarly, although they are less common in the uplands and prefer territories which include a stream or other open water. Wrens, too, commonly use walls as nest sites but, again, they are less common in upland areas, probably because of the colder winters there.

Walls, like fences, are used as vantage points by predatory species such as crows, but in upland areas there are also several smaller species which use such perches to search for insects, examples being Stonechat, Whinchat and indeed Wheatear. Especially in the uplands, a wall may be one of the taller structures available, and may be used extensively as hunting perches by all these species. Use by Carrion Crows and other predators will be detrimental to other species, as it is in the case of fences.

The size and shape of the wall seem to affect birds rather little. Birds will not use low and spread-out piles of stones as vantage points, such as where a long stretch of wall has fallen down, but such piles of fallen stones, and even a short stretch of them, are common nest sites for Wheatears and other hole nesters. In an otherwise well-maintained

Records of birds were scored along five parallel field boundaries across a farm in two years. The central boundary was a hedge (total 550 m), and the two outer ones on each side were post and wire fence lines (total 2100 m). In all, ten bird species were recorded along these boundaries with 101 records of eight species along the hedge (9.1 per 100 m per year) and forty one records of seven species along the fences (1.0 per 100 m per year). Along the fences, 49% of records were Corn Buntings, 20% Yellowhammers, both mainly of singing birds, and 20% were Carrion Crows. The remaining 11% of records were of Magpie,

Tree Sparrow, Greenfinch and Linnet. Along the hedge, 48% of records were of Yellowhammers, 21% Dunnocks, 10% Tree Sparrows, 8% Corn Buntings and 7% Blackbirds, while the remaining 6% comprised Carrion Crows, Magpies and Chaffinches together.

Carrion Crow and Corn Bunting were the only two species on this farm which used fences more than hedges, presumably as foraging vantage points and song-posts respectively.

Source: CBC files for a farm in Yorkshire.

---

wall, fallen piles of stones may indeed be the only suitable nest site. Therefore, in areas where Wheatears or Pied Wagtails are present, rebuilding a wall which has been fallen long enough for them to have selected it as a nest site (ten days or so) should be avoided between about mid-April and July.

Walls usually occur on their own. There may be a few bushes, especially in field corners, but only rarely will these be sufficient to make any difference to the quantity of birds. Hedges are present along some walls and in such cases the hedge, rather than the wall, is then usually the factor influencing bird numbers.

Walls are only used as nesting sites by a limited range of species, but are much used as vantage points by predators.

4.4 Banks

Many hedges, and especially boundaries which are grass strips, are on a slightly higher ridge than the field itself. These low ridges are often

referred to as 'banks'. However, in south-west England and Wales in particular, many field boundaries are more substantial earth banks, perhaps up to 2 m high and often with a hedge on the top. Such banks often contain rocks and stones and in many cases are derived from overgrown and earthed-in walls.

The advantage of a bank of any kind, compared to an unraised boundary, is better drainage. Hence any ground-nesting birds will more easily avoid flooding and waterlogging. This is probably the reason for banks being particularly favoured nest sites for the Grey Partridge, so long as there is sufficient grass cover to conceal the sitting female (ref 2 and see above). A similar preference is likely for other ground-nesting species, but there is no specific information. In the same way, a bank which faces south is likely to be warmer and drier than one facing other directions but, again, there is no specific information on whether or not these are used more.

Obviously, the large banks which are found in parts of western Britain provide more shelter than hedges or indeed stone walls. Occasionally the banks have suitable crevices for hole-nesting

Box 4.3 The use of stone walls by birds – an example

On about a third of the farm (23 ha) the field boundaries were drystone walls on their own – although this area included only about a quarter of the total field boundary length. On the remainder the fields were bounded by hedges in conjunction with walls, and with trees present in several boundaries. On this farm the 'walled' section was on the higher ground (320 to 350 m), and the 'hedge' section on the lower (270 to 320 m). The table shows the numbers of territories recorded on the two sections in four years combined.

Of these species, the Wheatear is a typical bird of stone wall country, and the Pied Wagtail also needs crevices for its nest. The presence of Blackbird and especially Linnet in the 'wall' section can be attributed to the presence of a few bushes in and around the fields, which those species were no doubt using. The Meadow Pipits and Skylarks probably preferred this area of the farm because the fields were rather larger and appeared more open as there were no trees in this section. These two species are known to prefer more open country (see ch 6).

	Number of territories in:		
	'Wall' section	'Hedge' section	Percentage on 'wall'
Skylark	16	3	84
Meadow Pipit	13	8	62
Pied Wagtail	3	6	33
Wheatear	3	1	75
Blackbird	1	36	3
Linnet	6	3	67
16 other species	0	209	0
Total	42	266	14

Source: CBC files for a farm in Derbyshire.

Box 4.4 Numbers of birds associated with ditches

On a series of 5 ha study sites on CBC farms in eastern and south-eastern England, Arnold[4] found that those sites which contained a ditch held twice as many individual birds as those without, and the average number of species was 3.4 compared to 1.5. Similarly in winter, on his 5 ha study sites, the total number of species recorded was 7.5 on those with a ditch and 5 on those without, and the number of birds seen per hectare was 2.7 with a ditch compared to 2.5 without. Study sites which also contained hedges and other woody vegetation held more individuals and more species in all cases.

On his particular study sites Arnold found that the volume of the ditch was a major significant factor in determining winter numbers of several species, in particular Blackbird, Dunnock, and buntings (Reed Bunting and Yellowhammer combined). In the breeding season the number of herb species in the ditch and hedgerow was one of the four most important variables for the total number of bird species recorded on the plot, and the numbers of individuals of some groups of birds[4].

On a Dorset farm an increase in the area of ditch bank associated with the hedge significantly increased the number of species present in both the breeding season[9] and in the winter[10], and the total number of individuals and the number of Blackbirds in the breeding season[9].

In Northern Ireland, 18 species were found to make significant use of wet ditches as feeding sites[11]. Snipe, Wren, Dunnock, Blackbird, Starling and Reed Bunting were prime users in winter, and in summer so were Song Thrushes. Summertime feeding in ditches was particularly common during periods of dry weather, when the ground elsewhere was quite hard and invertebrates would have burrowed deeper into the soil. In winter, ditches are also a favourite feeding site for Woodcock during cold weather[12].

Sources: Arnold[4], Osborne[9], Osborne & Osborne[10] and Moles[11].

birds, but the birds which occur along such banks are probably more dependent on the hedge or other woody vegetation present than on the bank itself.

Management of banks entails no particular problems, but the management of the associated vegetation is important and similar points apply as to such vegetation not on banks. See ch 3 for hedges, and section 4.1 for grass strips.

A raised bank is preferred as a nest site by Grey Partridges, and large banks provide shelter.

4.5 Ditches

There are two kinds of ditch, although they are not exclusive. Firstly, there is the type of ditch which is usually associated with a hedge and which is present mainly to drain off excess water. Although often dry, it may become a major channel for flowing water at certain seasons or after storms. Secondly, there is the permanently full ditch, or dyke, which is itself the main field boundary feature. The latter occur especially in more low-lying parts of the country, such as the Somerset Levels (where they are known as rhines), in the Fen District, and in coastal reclamation areas. These do serve as drains for excess water, but there is often little or no flow. From the birds' viewpoint they are very similar to ponds or slow-flowing streams or canals – see chapters 9 (ponds) and 10 (streams) for relevant details. This section here concentrates on the first type.

Hedges with a ditch alongside tend to hold more birds than do hedges without, and for some birds the ditch is one of the most attractive features. Some figures from published studies are given in Box 4.4

A deeper ditch will provide more shelter than a shallow one, but it is likely to be the greater abundance and availability of food which is the most important factor in attracting birds. Ground-feeding species seem to use the ditches most, and in particular species which dig or probe into the ground surface (see Box 4.4). Presumably this is because the soil on the sides and bottom of the ditch remains softer than in the surrounding areas, and especially in dry or frosty weather. In addition, softer ground is easier for invertebrates to move through, and so they are likely to be more available to birds. Earthworms are known to move lower within the soil profile during dry or frosty weather[13]. Surface water in the ditch will also attract insects of various kinds, which will provide additional potential food for birds. For these to be readily accessible, though, some shallow areas in the base of the ditch, rather than continuously steep sloping sides, will be particularly useful – see also ch 10.

Left unmanaged, vegetation in most ditches grows very fast and may quickly block the channel. Therefore management of ditches usually includes clearing away much of this vegetation, at least from the main channel. This may be essential to maintain the flow of water, and some birds, including some of those mentioned above, prefer to feed on fairly open ground. However, it is clear that some birds will occur only if there is some vegetation – see Arnold's study[4] noted in

Box 4.4. A careful compromise, involving the removal only of vegetation actually blocking the channel, is often the best overall solution.

The effects on birds of blanket vegetation removal can be fairly drastic, especially if it includes the removal of bushes and other woody vegetation. Two examples are given in Box 4.5.

The figures in Box 4.5 also illustrate a management practice which should always be adopted if possible. While the channel at water level will often need to be clear of vegetation, it is usually unnecessary to remove all the growth up the sides and especially on top of the banks. Machine clearance will require access, so that some destruction will probably be unavoidable. But it is usually unnecessary to clear vegetation from both sides, and much less damage is caused by working from one side only. The result will leave vegetation cover on one bank and some open ground on the other, and hence satisfy a variety of birds. Similarly, it is usually unnecessary to clear out the ditches every year, and clearing the water level vegetation from each side alternately is an attractive solution from farming, aesthetic and wildlife viewpoints.

Ditches are very attractive to several bird species as feeding sites, especially in dry weather and frosty conditions. Most birds prefer some vegetation to be left on the sides. Working from one side only when clearing will reduce potential habitat damage, and leaving some areas of shallow water at the edge of the channel may increase the numbers of birds able to feed there.

4.6 Summary and Management Recommendations

Grass strips

The birds which use grass strips for nesting (especially Grey Partridge and Skylark) and for feeding (especially Barn Owls) all prefer the

Box 4.5 The results of clearing two ditches

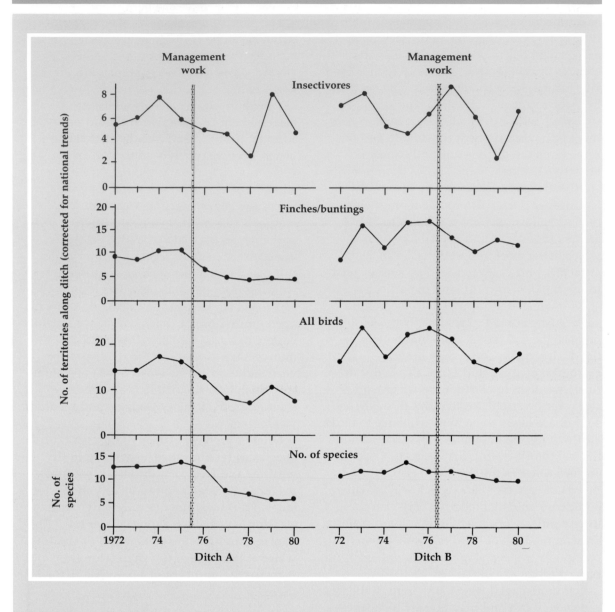

The diagram presents the numbers of territories and of territory-holding species and numbers of two individual groups of birds, along two ditches over seven years. The figures for number of territories are corrected to take account of national trends in numbers. The insectivores are Wren,

Dunnock, Robin, Blackbird and Song Thrush.

Ditch A: 400 m long. In the 1975/76 winter all the vegetation was removed from one side and most from the other, and the bottom was excavated.

grass to be at least 25–30 cm high. Species which habitually feed on short grass are unlikely to use strips. The largest amount of grass was found in strips managed in alternate years. A good practice therefore is to manage those on half the farm each year. Nesting species prefer strips to be wide enough to ensure sitting birds are not seen by passing predators, and Barn Owls need strips at least 5 m wide to ensure enough food.

Fences

Corn Buntings, and sometimes a few other species, may use fences as song posts, and one or two predators (Carrion Crow in particular) may use them as vantage points to look for prey. The last are detrimental to any other species present in the boundary or nearby in the field.

Walls

Walls are hardly used as nesting sites by birds, except by Wheatears and Pied Wagtails if there is a suitable crevice or hole. Wheatears and other chats may, however, use them as vantage points for prey searching, as may crows (see Fences).

Banks

Banks, as ridges even slightly above the level of the field, are especially favoured for nest sites by Grey Partridges – as long as there is sufficient grass or other vegetation cover – probably because they are better drained than flat glass strips. South-facing ones are likely to be preferred most, since they will be warmer.

Large banks offer more shelter than a hedge and often more than a wall, but otherwise such banks offer no special features for birds.

Ditches

Ditches are much used as feeding sites especially in dry conditions and during frosty weather. Most birds prefer some vegetation cover left alongside, and clearance reduces the numbers and variety of bird species using the ditch. When managing ditches, some benefits to birds can be achieved without loss of ditch function by working from one side only and by not clearing all the vegetation from both sides simultaneously. Clearing vegetation from each side alternately is often an acceptable compromise for drainage and birds.

Some shallower patches left on the edge of the water in the ditch may enable extra birds to feed on any invertebrates attracted. Permanent dykes are considered in chapter 10.

References

1 Potts, G.R. 1986. *The Partridge. Pesticides, Predation and Conservation.* Blackwell Scientific Publications, Oxford.

2 Rands, M.R.W. 1987. Hedgerow management for the conservation of partridges *Perdix perdix* and *Alectoris rufa. Biol. Conserv.* 40: 127–139.

3 Laursen, K. 1981. Birds on roadside verges and the effect of mowing on frequency and distribution. *Biol. Conserv.* 20: 59–68.

4 Arnold, G.W. 1983. The influence of ditch and hedgerow structure, length of hedgerows, and area of woodland and garden on bird numbers in farmland. *J. appl. Ecol.* 20: 731–750.

5 Schläpfer, A. 1988. Populationsökologie der Feldlerche *Alauda arvensis* in der intensiv genutzen Agrarlandschaft. *Orn. Beob.* 85: 309–371.

6 Shawyer, C.R. 1987. *The Barn Owl in the British Isles. Its Past, Present and Future.* The Hawk Trust, London.

7 Thomas, M. 1989. Causes of variation in the numbers of predators of cereal aphids overwintering in field boundaries. *The Game Conservancy Review of 1988:* 71–72.

8 Lack, P.C. 1986. *The Atlas of Wintering Birds in Britain and Ireland.* T. & A.D. Poyser, Calton.

9 Osborne, P.E. 1984. Bird numbers and habitat characteristics in farmland hedgerows. *J. appl. Ecol.* 21: 63–82.

10 Osborne, P.E. & Osborne, L. 1985. A winter bird census on farmland. Pp. 287–299 in Taylor, K., Fuller, R.J. & Lack, P.C. (eds). *Bird Census and Atlas Studies. Proc. VIII Int. Conf. Bird Census and Atlas Work.* British Trust for Ornithology, Tring.

11 Moles, R. 1975. *Wildlife diversity in relation to farming practice in County Down, N. Ireland.* Unpublished Ph.D. thesis, Queen's University, Belfast.

12 Personal observation, and G.J.M. Hirons personal communication.

13 Gerrard, B.M. 1967. Factors affecting earthworms in pastures. *J. Anim. Ecol.* 36: 235–252.

CHAPTER 5

Field Margins

The margins of fields are important because they form a link between the farmed land and the unfarmed areas such as field boundaries and woodland. Very often there is a boundary strip of some kind, and as such the margins can serve as buffer zones, preventing encroachment of undesirable elements from one to the other. On many farms the strip is very small or missing altogether, but alongside some fields there are tracks, grass strips, game crops or other types of vegetation or bare ground. All of these have implications for birds and other wildlife, and many of them are beneficial. Additionally, the edge of the crop itself may be rather different to the remainder of the field.

Over the last few years the Game Conservancy Trust has been carrying out specific research on this area of farms, in particular with their Cereals and Gamebirds Research Project. As a result it now promotes Conservation Headlands (see below) as a method of managing the field margins and headlands around cereal fields, with both shooting and conservation in mind.

This chapter considers both what is normally considered to be the boundary strip itself and the edge of the crop, although the latter may not be managed differently from the remainder of the field.

The commonest form of boundary strip is simply an extension of the boundary itself. Whether the boundary marker is a hedge, a wall, a fence or some other structure, there is often a strip of rank grasses or other herbs, maybe only a quarter to half a metre wide, between the base of it and the edge of the crop. Most arable farmers keep such a strip as narrow as possible, both because it takes up space which could be used to grow crops, and because it is often considered to be a source of weeds which might invade the crop. On

livestock farms, such a strip is not usually readily distinguishable as a separate entity unless cattle and sheep have been prevented from grazing right into the base of the hedge by for example a fence.

Even a narrow rough grass strip has some conservation value, but farmers can aid conservation more by some positive actions. Retaining existing features (e.g. tracks or putting in others (e.g. bare ground, or sowing wider grass strips) can be beneficial to several birds. Such strips usually have only a small cost in terms of lost production and may indeed have some direct economic benefits.

A barrier between crop and hedge is perceived by many farmers as important to prevent weeds encroaching from the hedge, although in most cases this is not justified[1]. The barrier also acts as a buffer zone, and as such can protect the base of the hedge from accidental, or even deliberate, spraying of fertilizers and pesticides beyond the edge of the crop. In these cases wide barriers are likely to be more effective than narrow ones, but even narrow ones have some advantages for birds. The different types are discussed below.

5.1 Management of existing narrow strips

A strip of rank grass alongside the base of a hedge is especially important if the base of the hedge itself is largely bare of ground-layer vegetation. It is also important alongside walls, fences, banks or other boundaries, for it may then be the only non-crop vegetation in the field.

Ground vegetation at the base of a hedge is vital for some birds. Grey Partridges in particular need such, preferably 1 m or more wide, for their nests, and other species, including Reed Bunting and Whitethroat, also nest there if there is some woody vegetation nearby. However a thick mat of grass and other ground vegetation will limit movement along the boundary and restrict access by feeding birds into the strip itself. This last is probably the reason why Grey Partridges are found to prefer such strips if they are cut in alternate years[2] – see also ch 3 and 4, especially Box 4.1.

In existing strips there is often less vegetation than there might be, especially in arable fields, because herbicides and other chemicals may be accidentally or deliberately sprayed into them. Further discussion of this, and additional management suggestions for such existing narrow strips, are given in ch 3.3 on management of hedges, and in ch 4.1 on grass strips as field boundaries. After all, narrow grass boundary strips are usually simply extensions of these.

5.2   Tracks and Farm Roads

A track is in effect a wide strip along the edge of a field, and usually contains grass or other herbage which is kept quite short. In addition, there are often bare patches of earth or mud, perhaps some puddles, and there may be sections with hardcore and even concrete or tarmac. Such 'roads' sometimes have a verge with rather longer vegetation.

From a bird's viewpoint, tracks are open areas of short vegetation or bare ground which are disturbed to a greater or lesser extent by farm vehicles. Because of this disturbance the track itself is unlikely to be useful for nesting sites, although the more luxuriant vegetation along the verge is often used by Skylarks[3]. There are no specific data on birds using tracks and roads for feeding, but it seems likely that those which come out of hedges onto grazed pasture fields, such as Blackbird, Robin and Dunnock, and perhaps finches will be the most common users. Birds which use the middle of large grass fields are much less likely to use tracks and roads as they prefer to be away from

hedges and other woody vegetation – see ch 6. Puddles may attract birds in dry weather, and House Martins and Swallows often use them as a source of mud for their nests. A few birds too will use dry bare patches for dust-bathing.

Traffic along tracks and roads beside fields can cause disturbance to birds nesting in the fields themselves. In the Netherlands, the densities of breeding Lapwing and Black-tailed Godwit were reduced within 400 m of tarred roads[4], and it was thought that disturbance caused by the passage of vehicles was largely responsible. A public road, as in this case, carries much more traffic than a farm track and hence will be a greater potential disturbance factor. But the effects of farm tracks on birds in the fields, such as Lapwings, should not be ignored. A track around the edge of a field is preferable to one crossing the middle, and screening a track with a hedge should further reduce disturbance.

The effects are not confined to nesting birds. Woodpigeons feeding in brassica fields were always disturbed if pedestrians, cyclists or horsemen went past on a road, but often remained feeding when a vehicle went past without stopping[5]. Such an effect probably applies to other birds feeding in fields too.

5.3   Bare Strips

Nearly a third of farmers now leave or create a strip of bare ground around their arable crops, in order to prevent encroachment of weeds from the hedge[1], although this is not as severe a problem as is often thought except for such as cleavers and sterile brome. Such a strip can be created in two ways: mechanically by cultivating, or chemically by applying an all-purpose herbicide such as atrazine. A mechanically created strip may have to be reworked once or twice during the growing season for it to remain free of ground vegetation and hence act as a barrier, but a chemical strip may well last throughout the growing season. Therefore the latter is likely to

Box 5.1 Use of bare strips as feeding sites by hedgerow birds

The diagram (adapted from Cracknell[6]) shows the use of a bare strip at the edge of a cereal field compared to the use of the adjacent crop by four bird species on Manydown Farm in Hampshire. The figures are the number of visits made by individual birds out of the adjacent hedge in five minute periods of observation. Three of the four species visited the bare strip more than the adjacent headland, the far crop or the vegetation in the base of the hedge, and if the area of each is taken into account, all four visited the bare strip considerably more often than the rest put together. (The bare strip was a much smaller area than the others except for the

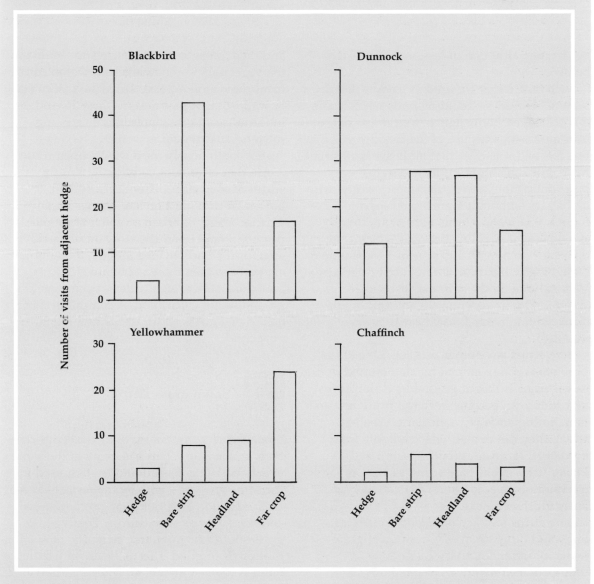

hedge base vegetation.) The reason was presumably that the birds were able to move around and find food more easily than in the crop.

Spott[7] compared field boundaries with and without bare strips between the crop and the hedge. More birds came out more often onto bare strips even though the hedges in this part of the study farm were smaller and probably contained fewer birds overall than the part of the farm where strips were absent.

be cheaper, although of less use to birds (see below).

Such bare strips are used as feeding sites by a variety of species, including birds which come out of hedges. Some figures for the use of them in comparison with use of the crop are given in Box 5.1, and it is clear that birds use bare strips more. For such birds, cultivated strips are potentially more useful than those created with herbicides, although there are no specific figures. Cultivation turns the soil and often exposes seeds and invertebrates which were invisible. Obviously, any repeated rotovation or ploughing will expose a new supply, perhaps at a critical time in the summer when soil invertebrates are deep in the soil and out of reach, having moved downwards in dry weather.

Bare strips have other uses too. One of the problems for any animal moving around in cereal crops is that at ground level the leaves and stems can become very wet from dew and rain. Young birds in particular need areas where they can dry off which are safe from predators. Partridges (both species) and Pheasants will use bare strips, as well as tracks or short grass areas, as relatively safe areas for their chicks.

Bare strips also act as corridors enabling birds and other animals to move along the hedge or other boundary easily and unimpeded. Hedgerow-dwelling birds are seen regularly flying low along these strips, and there seem to be advantages for birds in the vegetation in the hedge base. For example, foxes moving along bare strips have been seen to walk straight past nests of Grey Partridges in the adjacent grass, presumably missing them because the foxes were not having to search continuously for the most open route.

Another important consideration is the width of such strips. A wider strip will obviously provide a greater barrier for any encroachment. Similarly, a wider strip will provide a larger and therefore more useful area for the birds to feed upon, although they are perhaps then more vulnerable to predators. However, a wider strip means less ground under cultivation. A compromise of half to one metre width seems to serve most purposes.

5.4 Sown grass strips

Some farmers sow a strip of grass alongside their arable fields. This has many of the same effects as the bare strips and is often used for the same purposes, both by the farmers and by the birds. However, there are two additional choices if a grass strip is used. Firstly what grass species to grow and, secondly, how tall it is allowed to grow. This depends on whether or not it is mowed and, if so, when and how often.

For birds, the species of grass or herb is not

usually important although the seeds of some are a food supply for finches and buntings in the autumn.

As noted above, both partridge species, and some others, prefer quite long grass for nesting, at least in the base of the hedge. However, most species are unlikely to nest more than 1 m or so away from the hedge, so other considerations are probably more important for grass as a boundary strip. Most small hedgerow birds which feed on the ground prefer short grass to that which is tall (see ch 4). Grazing, and hence temporary fencing, are usually impractical alongside arable crops, but mowing is practicable and many farmers who sow a grass strip also mow it once or twice per year. This serves the dual purpose of removing any plants which would be undesirable in the fields, such as cleavers and sterile brome, and creates a shorter sward which may be used by birds just as if it was a bare strip.

However, Barn Owls prefer grass strips which have been only lightly grazed, and certainly not the short grass preferred by most other bird species – see ch 4.1 for a fuller discussion. For wider strips, there is of course a possible compromise. Some parts can be mown and kept short and some parts can be allowed to grow. For many reasons the taller, grass is likely to be best situated adjacent to the hedge, and the shorter grass adjacent to the crop.

5.5 Game Crops

Game crops are usually planted only on farms where there is an established shoot. Common species used include various legumes and root crops, and maize, although there are several others. The Game Conservancy Trust now produces a 'cocktail game mixture' of sunflower, maize, canary seed, buckwheat, caraway, American sweet clover, marrowstem kale and fodder rape as a good all-purpose game crop for use in southern England, although the authors of a recent book on game crops[8] point out that in other areas and on less

common soil types some modifications are likely to prove more suitable.

The main purpose of a game crop is to increase potential nesting and holding cover for gamebirds, especially Pheasants, to draw birds into a single area for improved shooting, and to provide extra food. For other birds, an established game crop will always provide some cover either for nesting or for shelter in the winter or both, and often it provides food too. Gamebirds eat green vegetation as well as seeds, and their chicks eat insects in the spring and summer, but for other birds the attraction is primarily the seeds and insects. Finches and buntings in particular will use game crops in the autumn and winter, and they may lead to a concentration of such birds, especially in areas with little or no other potential food.

For shooting, the siting of the crops is critical to providing sporting birds for the guns, and the Game Conservancy advisory booklet[8] covers the subject thoroughly. A game crop is most likely to be used by other birds when it is sited close to some woody cover, for most birds prefer to be able to retreat into cover when disturbed. This includes the finch flocks, although flocks are prepared to move further out into fields than are single birds.

For more details of the advantages and disadvantages of different plant species, and of all other aspects related to the planting and management of game crops with respect to shooting, see the Game Conservancy booklet[8].

5.6 Conservation Headlands

The Game Conservancy Trust has recently been promoting what it terms Conservation Headlands as a means of increasing the productivity of gamebirds in arable farmland, and in the process increasing the numbers and diversity of some other groups of animals and plants. Conservation Headlands are the outer few metres (normally the 6 m from the edge of the field to the first tramline all around the field) which are subjected to different management than the rest of the field, especially with respect to the use of pesticides. Specific recommendations are available from the Game Conservancy Trust. Essentially, the Conservation Headland is treated with many fewer pesticides. Insecticides should not be used after 15th March on autumn cereals, and none at all should be used on spring cereals. A limited range of herbicides may be used sparingly, with broad-leaved weed killers being subject to particular restrictions, but fungicides may be used except for those which are known to have insecticidal properties.

Game Conservancy Trust research has shown that treating headlands in this way produces a field edge with many broad-leaved plants and insects and, in most cases, there is only a very

Box 5.2 The use of conventional and conservation headlands by hedgerow birds

The diagram illustrates the numbers of feeding visits to conventionally managed headlands and Conservation Headlands by four bird species. Clearly there was no difference in the use of the two types by each species or all species combined. Indeed, in some cases, there was actually more activity in the conventionally managed headlands. This was probably because such areas contained less vegetation at ground level so that birds found it easier to move around.

Censuses indicated that there were no significant differences in numbers of birds along hedges adjacent to the two types[14]. Similarly, there were no indications of greater or lesser breeding success, or of more or less movement, around the hedgerows adjacent to the two types, although the samples for these were very small[6]. However, in interpreting these figures the cautions about the timing and placing of the conservation headlands on this study site noted on p. 53 should be taken into account.

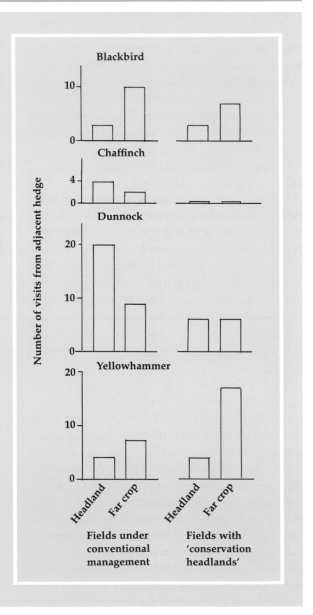

Sources: Fuller[14] and Cracknell[6].

small effect on overall crop yield from the field[9]. For example, a square field of 20 ha with a 6 m headland all around is unlikely to show a yield reduction of more than 0.5%, and a 10 ha field more than 0.75%. However, particular weed species can sometimes cause more severe problems.

The major reason for creating Conservation Headlands is that there are large benefits for gamebirds. The survival of partridge chicks is doubled[10] and that of Pheasant chicks increased substantially[11]. In addition, butterflies are seen more often over such headlands than over conventionally managed crop edges[12], small mammals use the areas in preference to conventional headlands[13], and some plants of conservation interest are enabled to survive. However, the use of Conservation Headlands by birds other than gamebirds, and particularly by those in the adjacent hedgerows, seems to be minimal and not significantly different from the use of normally sprayed headlands[6,14]. Some figures are in Box 5.2. Birds which typically feed in the middle of fields usually prefer to feed away from the hedges and edges of the field, so are unlikely to be much affected by different management strategies on the headlands.

Two cautions ought to be raised, though, about the interpretation of Conservation Headlands being of no special interest to birds other than game birds. The first is that hedgerow birds do not use adjacent arable, and especially cereal, fields very much at all as feeding sites; and second, in the studies carried out on the Manydown Estate[6,14] the bird species concerned will have set up their territories in the hedges before any differences between fields became obvious in the spring. During the experimental phase of the Cereals and Gamebirds Research Project, when this bird work was done, Conservation Headlands were placed around different fields each year, so there was no history of more plants and insects in one place rather than another. However, since then the individual fields have tended to be permanently under the regime, so it would be worth repeating the observations on a longer term basis. The

argument that birds find it more difficult to move about on the ground in Conservation Headlands will hold wherever such headlands are placed, but it seems intuitively unlikely that birds will avoid an area with many insects in it unless there is a major access difficulty.

5.7 Summary and Management Recommendations

General

Boundary strips serve to separate the crop from the boundary and, as such, are a useful barrier for preventing encroachment of the crop by potential weeds (although this is often overstated as a problem) and preventing harmful sprays getting into a vulnerable hedge bottom.

Existing narrow strips

Even a narrow strip of rank grass and other herbs will often be used for nesting by partridges and a few other bird species. Cutting in alternate years has been found to be best for partridge survival. See also ch 3.3 and 4.1.

Tracks

Tracks and their verges provide feeding sites for hedgerow birds and they may act as corridors. A thickly vegetated verge may be used as a nest site, by Skylarks for example.

Traffic will cause some disturbance to birds in a field, but a screen will reduce this considerably. Tracks across the middle of fields will cause much more disturbance than those around the edge.

Bare strips

These are used by several bird species for feeding, by some mammals as a path, and by gamebird chicks as an area in which to dry off. Strips can be created by cultivation, which may need to be done two or more times each growing season, or chemically with a herbicide such as atrazine, which will probably only need to be applied once. Cultivation of strips may serve to renew a food supply of seeds or invertebrates and hence is of greater potential use to birds.

Sown grass strips

These may be used by several bird species for feeding and have many of the characteristics of bare ground strips. However, they can be left unmown and they may then provide feeding areas for Barn Owls. With a wider strip the part nearer the crop can be kept short by mowing, and that nearer the hedge longer to provide cover for nests or potential prey for owls.

Game crops

These are planted as cover and food, especially for Pheasants, but they are often used by finch flocks in the autumn. Siting is critical for providing sporting birds for shoots, but neither this nor plant species seem to be very critical for other bird species. See the Game Conservancy Trust booklet (ref. 8) for more details.

Conservation Headlands (as defined by the Game Conservancy Trust)

These double the survival of Grey Partridge chicks, increase the survival of Pheasant chicks, increase the numbers of butterflies seen, increase the numbers of small mammals which use the field, and the numbers of some plants present, but seem to have rather little effect on small birds in the hedges. However, hedgerow birds do not use arable fields or headlands very much anyway.

References

1 Marshall, E.J.P. & Smith, B.D. 1987. Field margin flora and fauna; interaction with agriculture. Pp. 23–33 in *Field Margins* (eds J.M. Way & P.W. Greig-Smith). British Crop Protection Council Monograph no. 35, Thornton Heath.

2 Rands, M.R.W. 1987. Hedgerow management for the conservation of partridges *Perdix perdix* and *Alectoris rufa*. *Biol. Conserv.* 40: 127–139.

3 Laursen, K. 1981. Birds on roadside verges and the effect of mowing on frequency and distribution. *Biol. Conserv.* 20: 59–68.

4 van der Zande, A.N., ter Kewis, W.J. & van der Weijden, W.J. 1980. The impact of roads on the densities of four bird species in an open field habitat—evidence of a long distance effect. *Biol. Conserv.* 18: 299–321.

5 Kenward, R.E. 1978. The influence of human and goshawk (*Accipiter gentilis*) activity on woodpigeons (*Columba palumbus*) at brassica feeding sites. *Ann. appl. Biol.* 89: 277–286.

6 Cracknell, G.S. 1986. *The effects on songbirds of leaving cereal crop headlands unsprayed.* A report to the Game Conservancy. BTO Research Report no. 18. British Trust for Ornithology, Tring.

7 Spott, C. 1989. *The importance of crops as foraging sites for hedgerow birds.* Unpublished thesis for Diplomarbeit, University of Wurzburg.

8 Game Conservancy. 1986. *Game and Shooting Crops.* The Game Conservancy, Fordingbridge.

9 Boatman, N.D. & Sotherton, N.W. 1988. The agronomic consequences and costs of managing field margins for game and wildlife conservation. *Aspects of Appl. Biol.* 17: 47–56.

10 Rands, M.R.W. 1985. Pesticide use on cereals and the survival of grey partridge chicks: a field experiment. *J. appl. Ecol.* 22: 49–54.

11 Rands, M.R.W. 1986. The survival of gamebird (Galliformes) chicks in relation to pesticide use on cereals. *Ibis* 128: 57–64.

12 Rands, M.R.W. & Sotherton, N.W. 1986. Pesticide use on cereal crops and changes in the abundance of butterflies on arable farmland. *Biol. Conserv.* 36: 71–82.

13 Tew, T. 1989. The effects on small mammals of differing pesticide use on cereal field margins: a summary of three years' work. *The Game Conservancy Review of 1988*: 56–58.

14 Fuller, R.J. 1984. *The distribution and feeding behaviour of breeding songbirds on cereal farmland at Manydown Farm, Hampshire, in 1984.* A report to the Game Conservancy. BTO Research Report no. 16. British Trust for Ornithology, Tring.

CHAPTER 6

Fields and Crops

Fields and crops comprise the largest part of a farm and, unlike most other habitat types, they are always under active management. All agricultural practices in fields will affect some birds, while some of the practices adopted in the fields also have implications for birds in other habitat features.

This chapter indicates the main effects of agriculture on the birds which nest or feed in fields. Relatively few species place their nests there, but these include some of the rarer farmland birds and some, such as Stone Curlew and Black-tailed Godwit, which are attracting considerable attention from conservation bodies. Rather more species feed in the fields, especially outside the breeding season. Some of these are also of interest to conservation bodies, especially since quite large numbers of birds may be involved, and some species attract considerable interest from farmers as they may be pests in some circumstances.

The chapter points out what different bird species need from the fields and how different management practices will be beneficial or detrimental to them. Depending partly on what temporal or regional subsidies are available, some of the suggestions for changes which benefit birds will undoubtedly result in a reduced profit in purely monetary terms. At least some farmers will therefore consider them unacceptable. They are nevertheless included, for practices even part way towards the birds' ideal will often have some benefits. Other suggestions have only small or even no costs in economic terms, and could be incorporated fairly easily into the farming routine. In addition, some of the initiatives which have been suggested in recent years, such as extensification and set-aside, may have implications.

This chapter is concerned with the field as a cropping entity. The margins, where there are often other and specific aspects to consider, have been dealt with in chapter 5.

6.1 Use of different field types by birds

For nest sites by commoner species

The commonest and most widespread bird species nesting in the fields is the Skylark. The Common Birds Census data indicate that it is much more common in eastern England than western[1], and there is some evidence from individual farms within the CBC that, on mixed farms, it prefers to nest in arable crops rather than on permanent grass. Also, it prefers large fields surrounded by few and low hedges to small, tightly enclosed fields with trees scattered along the hedgerows. A detailed study in Switzerland[2] found, in addition, that it preferred farms containing small plots of a variety of crop types to those with large areas of monoculture. Winter cereals and oilseed rape were used quite extensively in April, but by June these were avoided because the vegetation was too high. The preference then turned especially to spring cereals, with road verges, fallow land and some other crops sometimes as important secondary habitats.

Only two others of the common breeding species of Box 1.1 (p. 4) are field nesting birds – the Lapwing and Yellow Wagtail. The Lapwing has been the subject of several recent studies, and Box 6.1 contains some figures on its preferences for different field types. The ideal seems to be a spring-sown arable crop for a nest site near to a grass field which is the ideal habitat for rearing chicks.

The Yellow Wagtail nests especially in tussocks in the wetter grass fields[5]. However,

Box 6.1 The field type preferences of Lapwings in the breeding season

A BTO survey of England and Wales estimated that just over 123,000 pairs of Lapwings nested in England and Wales in spring 1987[3]. The diagram here indicates the preference for nesting in each of five major field types in three large regions of England and Wales. (The index is calculated allowing for the availability of the different crop types. 'North' is the North, Northwest and Yorkshire/ Humberside regions combined; 'East' is the East Midlands, East Anglia and Southeast regions; and 'West' is the West Midlands, Southwest and Welsh regions, with the regions being the standard statistical regions as used by MAFF and others.)

Lapwings clearly avoid autumn sown crops, show a strong preference for spring sown crops, avoid leys and improved, ungrazed, permanent grass, and show a slight preference for rough grass.

It seems that the main requirement for nesting is fairly open vegetation so that the incubating birds can see any approaching potential predators. Nesting birds seem to avoid fields which contain dense vegetation in April, whether this is an arable crop or grass.

However, it is more complicated than this. The birds are more choosy still. In the BTO survey it was seen that spring sown crops, including vegetables, were especially favoured when there were grass fields adjacent to them[3]. It appears that the birds prefer to raise their chicks in grass fields[4], and especially those in which the grass is high enough to conceal the young birds from predators. The ideal for breeding Lapwings therefore seems to be a landscape in which there is both spring tilled arable crops and grass fields, preferably closely intermixed.

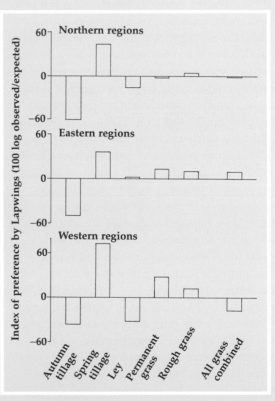

Sources: mainly the BTO survey of 1987[3] and Galbraith[4].

for feeding it also needs short grass or bare ground nearby. It also nests in arable fields, apparently without any major preferences for crop type (but see below).

A few other, more common, species nest occasionally in fields, examples being the Reed Bunting and Corn Bunting. Both of these often nest rather late in the season and seem to move out into the crops only when these are quite well grown. The Corn Bunting particularly favours barley fields[1,6]. The Reed Bunting will nest in barley[7], but there has recently been a strong preference for oilseed rape (see below). The occasional pair of Sedge Warbler and Dunnock also nests in rape fields later in the season (CBC data).

One means of finding out the effects of different crop types is to investigate farms where major changes have occurred, and

Box 6.2 gives some examples. It appears that common (mainly hedgerow) species are relatively unaffected, probably because they use the crops rather little, although they do show some preferences for feeding sites (see below). However some specialist and scarcer species are affected much more.

For nest sites by rarer species

Scarcer nesting species in lowland fields include the Meadow Pipit, which is mainly an upland bird, and some other waders, in particular Redshank, Curlew and Snipe. All these are primarily (if not entirely) grassland birds and all, especially the waders, prefer the wetter rough grass fields[9]. Wildfowl too prefer to nest in wet rough grass fields rather than other types. Further details of the preferences of waders and how management can be adapted

Box 6.2 The effects of changing farming type

During their census period several CBC farms have switched their major farming type. Of these, four which changed from under 50% to over 70% arable, and three which changed from under 40% to over 80% grass, were investigated over three years on either side of the change. The diagram on p. 58 gives the average numbers of birds recorded over the seven farms combined for four individual species, for all species combined and for the total number of species recorded.

There are indications that there are fewer species and lower overall numbers when the farms have larger areas of arable crops. (The decline in both total numbers and number of species is statistically significant.) There were no significant trends for any individual species, but there is a suggestion that Lapwings are commoner when there is more arable, and Blackbirds when there is more grass. This is the direction which could be expected from

other knowledge of their field type preferences (see Box 6.1 for Lapwing and Box 6.4 for Blackbird).

Unfortunately, very few of the rarer species occurred on any of these seven farms.

In Denmark[8] a progressive change over seven years from 100% grass to 70% cereals on an area of 14.66 km[2] led to decreases of 80% in numbers of Yellow Wagtails (12 to 2 pairs), of 65% in Lapwing (57 to 20) in contrast to the above, of 75% in Snipe (19 to 5), of 12% in Whinchat (26 to 23), of 60% in Meadow Pipit (19 to 8) and of 40% in Reed Bunting (61 to 37 pairs). In all these cases the decrease in bird numbers was thought to be due largely to the simple change from grass to cereals in the fields, although there were some other changes in the landscape at the same time, in particular drainage.

(diagram over page)

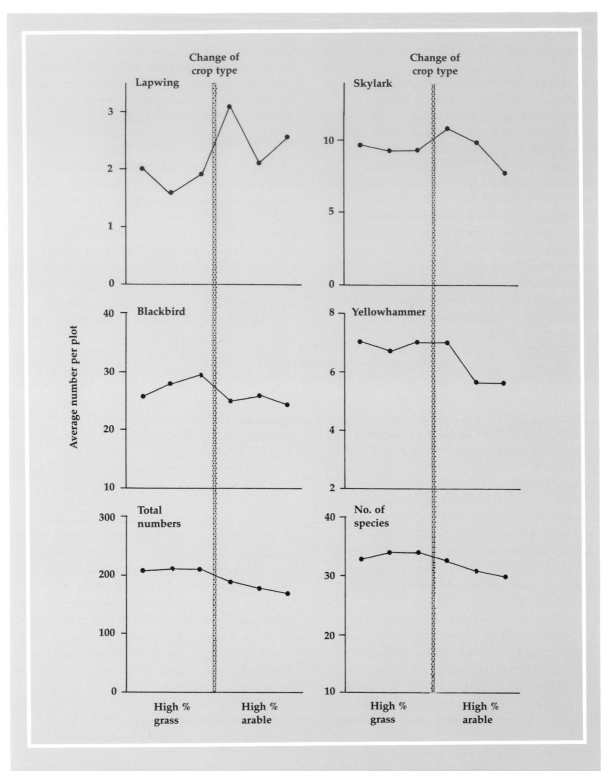

Sources: CBC data and Møller[8].

Box 6.3 Birds emerging from hedgerows onto different crop types

Arable:Grass Spott[13] compared a Hertfordshire cereal farm (entirely wheat) with an adjacent pasture farm. The hedges on both farms were fairly similar, although the total number of birds present was higher on the pasture farm. Blackbirds and Robins emerged from the hedges more commonly onto pasture fields, Yellowhammers, Chaffinches and Greenfinches more commonly into wheat fields, and Dunnocks about equally onto each type. However, the overall number of observations was rather small, and the relative use of pasture by all species was reduced if the number of breeding territories along adjacent hedges was taken into account. In addition to the above species, some Carrion Crows, Magpies, Starlings and a few others were seen feeding in the pasture fields during the study, although not seen emerging from the hedgerows. None of these or any other species was seen similarly in the wheat fields.

Wheat:Barley Comparisons of wheat and barley were made as part of the study of Conservation Headlands on Manydown Farm in Hampshire[14,15], although in neither case were the hedges standardised fully. The four commonest species seen to emerge from the hedgerows, Dunnock, Yellowhammer, Blackbird and Chaffinch, all fed in wheat more often than in barley fields[15]. For the same species there was no difference between use of spring sown barley and a combined figure for winter wheat and winter barley[14].

Other arable crops More individuals and species were seen feeding in beans, rape, and especially peas, than in wheat[16].

Tree Sparrows studied at the Boxworth Experimental farm in Cambridgeshire were found to visit winter wheat on about half the foraging trips out from nests, with other visits recorded to field beans, hedges, ditches, woods and gardens[17].

Sources: mainly BTO studies by Spott[13], Fuller[14] and Cracknell[15].

for their benefit are given below (p. 69) and in ref 10.

Harriers too, both Marsh and especially Montagu's, nest in fields. These prefer much longer vegetation than most other species and will often nest in well grown cereal fields[11]. These large raptors are amongst the rarest of British breeding birds. Another less common species which occurs mainly in well-grown cereal fields is the Quail. This species arrives usually in late May or June and numbers vary considerably between years.

In East Anglia, special interest centres on the Stone Curlew. This prefers to nest on sparsely vegetated ground. In Britain it is mainly on grassland which is heavily grazed by livestock or rabbits, and on spring tilled fields, i.e. much

like the Lapwing. The reason appears to be the same[12].

In summary, the field types most preferred by nesting birds are lightly grazed, wetter, rough grass fields (especially by waders), and combinations of spring sown tillage crops and grazed grass (e.g. Lapwings and Stone Curlews), although a few will nest in most crop types at times.

As feeding sites during the breeding season

The use of crops as feeding sites by birds has received remarkably little attention from scientists. However, obtaining unbiased information is complicated because birds tend to feed near their nest sites, and their choice of breeding territory may therefore be determined

Box 6.4 Preferences by birds for different arable crops for feeding

Records of birds actually in the fields were scored on the CBC maps of each of eight farms over at least four years. Most of these records would be of feeding birds although this was not usually stated explicitly. The number of records in each major crop type was then compared to the number expected if the crops had been used in proportion to their availability on the farm.

The preference indices of five species and of hedgerow passerines combined (but excluding Reed Bunting) for five main arable crop types are shown in the diagram. The five crop types were oilseed rape, wheat, barley, root crops and others – the last including hay, vegetables and unidentified cereals. Unfortunately it was not possible to distinguish autumn sown from spring sown crops in this analysis.

Among the field-nesting species, the Lapwing is seen to avoid oilseed rape and wheat, nearly all of which are autumn sown, and if anything to favour the others which will include some spring sown crops. Therefore this analysis agrees with the results of the special survey of this species (Box 6.1). The Yellow Wagtail showed a strong preference for root crops. The Skylark (not illustrated) showed no marked preferences in this analysis although oilseed rape was the least used crop.

Among hedgerow-nesting species there was a marked preference by several for oilseed rape fields, a marked avoidance of wheat and barley fields, and a rather mixed response to fields of roots and other crops. In contrast, the Carrion Crow showed a very similar pattern to the Yellow Wagtail.

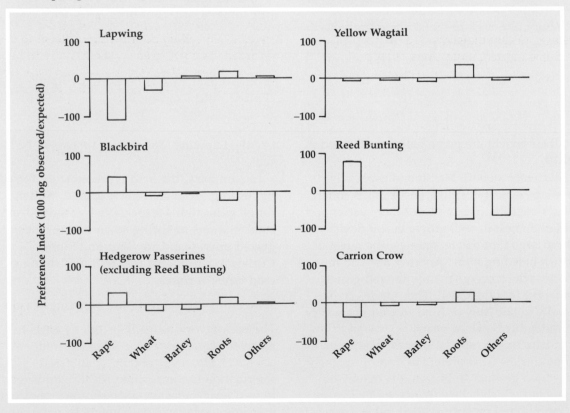

A few other particular preferences were noted in this analysis although not illustrated, including the Corn Bunting's preference for barley found elsewhere[1,10].

These preferences can best be explained by the accessibility and ease of moving around within the different crops. Oilseed rape and root crops are fairly open at ground level, although entering in the first place may be more difficult for birds. Cereals are much denser at ground level and this may prevent birds moving around in these fields.

Source: CBC data.

more by nesting requirements than by potential feeding sites adjacent to the nest. For example, hedges around pasture fields are often larger and thicker than those around arable crops. Therefore more birds may choose hedges around pasture fields, resulting in their use of these fields being determined by hedge quality rather than a preference for the field type.

Three BTO studies have investigated bird emergence from hedgerows on to fields. Details are presented in Box 6.3. Pasture seems to be preferred to arable crops, broad-leaved crops preferred to cereals and wheat preferred to barley. The reason in each case is likely to be related to ease of access into the crop.

Some indirect indications of crop preferences can be derived from CBC maps. Figures from some analyses of this are in Box 6.4.

On grassland there appears to be a marked preference for shorter grass, whether cut or grazed. Casual observations suggest that grazed pasture is used quite extensively by some birds, notably Carrion Crow, Magpie and Blackbird. Newly mown grass fields, for either hay or silage, are also heavily used by these and several other species (pers. obs.), and Mistle Thrushes are more common in areas where the proportion of mown fields is higher[1].

Feeding behaviour on grassland has been studied especially on airfields because of the importance of trying to keep birds away from aircraft. Over the whole year Lapwings, gulls, pigeons, crows and Starlings were found to occur on short grass (5 cm) sites five times more frequently than on long (15–20 cm) grass sites, and that short grass held three times as many birds overall[18]. Again the cause is probably the physical ease of access, although in this case too the food supply may play a part. Arthropod density has been found to be 30% higher in cut than in uncut grass although biomass was lower[19].

In summary, some bird species at least show individual preferences for different crops, but a few generalizations can be made. For example there seems to be a marked preference by hedgerow birds for grazed grass and broad leaved crops, especially oilseed rape; while long grass and cereals, especially autumn sown, seem to be avoided. The preferences seem likely to be due to the physical structure of the crops and the subsequent difficulty of moving into and within them.

As feeding and roosting sites in the non-breeding season

More birds use fields in winter than in the breeding season. Many of them occur in flocks and include migrants which come to Britain from breeding areas to the east and north. Residents include several crow species and migrants include Lapwing, Golden Plover, several thrushes (especially Fieldfare and Redwing), various gulls (particularly Black-headed) and Starlings. (Several of these have resident populations as well.) All of these feed mainly on invertebrates although the thrushes and Starling also eat many berries in hedges and scrub during the autumn. Rather few species feeding in fields in winter are vegetarian, but these include the finches, pigeons and geese.

Some figures for crop preferences by several invertebrate-feeding species are given in Box 6.5. Clearly, grass and stubble are the two most preferred field types by most species.

As in the breeding season it is the Lapwing which has been the subject of the most detailed study in the winter. In a site centred around an airfield Lapwings were found to prefer ploughed land, broken concrete and, especially, mown grassland[22], with the last for a longer period than either of the other field types. In central England, the winter density of Lapwings was higher on grassland than arable although ploughed land was especially favoured for roosting sites[23]. In Sussex, Lapwings did use ploughed land for roosting but preferred old tussocky grassland, especially the wetter fields[24]. The Golden Plover similarly preferred to feed on permanent pasture fields throughout the winter although it roosted mainly on ploughed land and turned more to autumn-sown cereal fields and ploughed land for feeding in the late winter[23,25]. Food

availability in the different field types is the most important factor, and permanent pasture holds more invertebrates than other types[20], but shelter is also presumed to be important for roosting and this is well provided by both tussocks and plough furrows.

Other species which occur in large flocks in winter are Starlings, Rooks and Jackdaws. These too are most common on permanent grass fields but, because they eat grain and seeds as well as invertebrates, they are relatively more common than plovers or thrushes on stubble fields.

Of the vegetarians, Woodpigeons and Skylarks seem to favour oilseed rape (BTO data) and sugar beet[26]. Woodpigeons can become a pest at times and have to be kept off with bird scarers or by shooting. A similar pattern can arise with geese in a few areas but, with the exception of the introduced Canada Goose, the distribution of geese in England and Wales is fairly restricted to a few sites and problems only arise locally[27,28]. Problems with Brent Geese in Essex and Sussex, and some suggested solutions, are discussed in Box 6.6, and other species and

Box 6.5 The use of different field types by birds in winter

The diagram presents the preferences of a range of species for different field types in the period November to February. The six types studied were stubble, bare tilled, winter cereal, oilseed rape, ley and permanent grass, and the species codes are L=Lapwing, BH=Black-headed Gull, FF=Fieldfare, MG=Magpie, JD=Jackdaw, C=Carrion Crow, and SG=Starling. The other birds studied were Golden Plover which showed a very similar pattern to the Lapwing, Rook which was very similar to the Jackdaw, and Blackbird, Song Thrush and Redwing which were all very similar to the Fieldfare.

It is clear that grass was the most preferred crop type overall and by most individual species. The other important type was stubble, although its importance for birds is considerably understated in this diagram, because it only includes invertebrate feeders. Stock Doves and Woodpigeons are quite commonly found on stubble, any many finches, buntings and sparrows also feed in this field type during the winter[21].

Source: BTO study by Tucker[20].

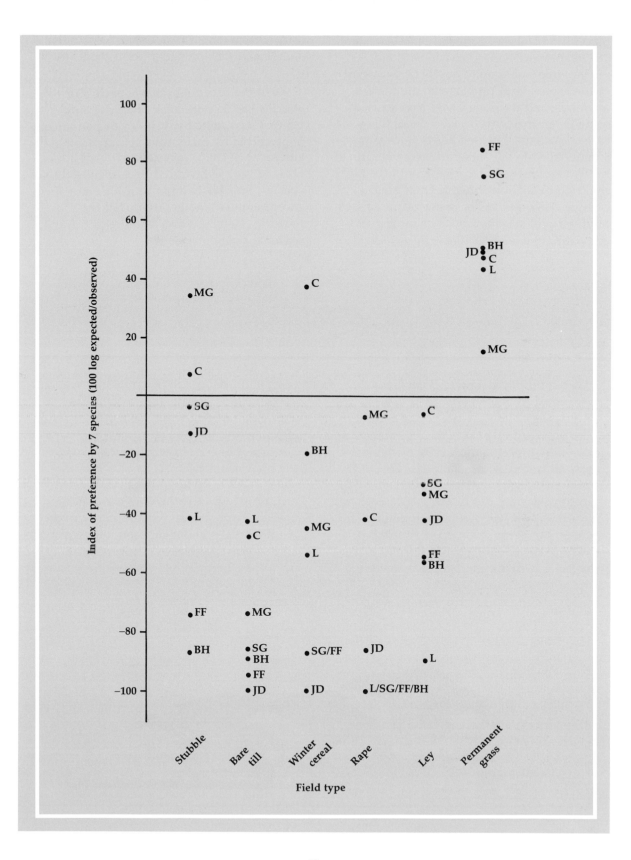

places are similar to a greater or lesser extent. Geese are more widespread in Scotland, where Pink-footed and Greylag Geese have created some local problems in the eastern lowlands, and Barnacle Geese have caused some major incidents on Islay. In all these cases the damage caused has been minimal on a national scale but sometimes very severe on individual farms, usually because of the birds selecting to feed on valuable crops such as carrots. The problems have been summarised briefly[28], and specific advice on means of alleviating them are available in advisory leaflets from the Scottish Office Agriculture and Fisheries Department[29] and the RSPB[30].

As in the breeding season, each bird species has its particular favourite crop types. Grass, whether it is ley or permanent, is preferred by many invertebrate feeders and some of the vegetarians also eat grass. Oilseed rape is favoured by some vegetarian species. Stubble is important for many birds and ploughed land is often used for roosting. There are also some regional and seasonal variations.

Box 6.6 The Brent Goose and farming

The world population of Brent Geese has increased from about 25,000 in 1960 to nearly 200,000 in the late 1980s, and they have increasingly taken to feeding inland on fields to supplement their traditional estuarine and saltmarsh foods. The preferred farm crops are barley and improved grass but they also graze wheat, grass leys and oilseed rape among others. Geese can account for up to 20% loss of yield locally, although they usually graze only a few fields even in problem coastal areas of Essex and Sussex.

Several methods have been tried to prevent or reduce the potential damage. These include scaring and shooting, but an alternative method which has proved very effective is to provide refuges (Alternative Feeding Areas). For these, grass is cut to the preferred height (about 5–10 cm) and some decoy birds are put out. If birds are also scared off other fields, a concentration of birds can be achieved quite easily in the required place, and away from vulnerable crops so that damage is much reduced.

Source: mainly Prater[31].

Farming operations and their timing

For birds, the two most important farming operations in the fields are cultivation of the ground and the harvest. Both of these remove one potential food supply and replace it with another, and both cause disturbance during the operation itself. Some other activities, including rolling, harrowing, spraying and manure spreading, also cause disturbance and may affect food supplies as well.

Ploughing and cultivating

A tractor ploughing a field may be followed by dozens or even a few hundred birds, notably Black-headed Gulls, Rooks, Starlings and sometimes Lapwings. All these birds are attracted by the opportunity to feed on the soil-living invertebrates which are brought to the surface by the plough, but these new food supplies only persist for a day or two after ploughing. Ploughed fields are also used widely as winter roosting sites by Lapwings and Golden Plovers (see above).

Yet there is a detrimental effect of ploughing, as it is most often of stubble fields. As noted earlier, stubbles are very attractive to birds, especially the seed-eaters. In the past, stubble was left through the winter and provided an important source of seeds, including spilt grain, for many bird species. By the spring most seeds would have been eaten, and ploughing would then renew the supply of accessible invertebrates at a time (February and March) when many birds experience some difficulty in finding food. With cereals now largely being sown in the autumn, fields rarely remain as stubble for more than a week or two after the harvest of the previous crop. Only in more northern areas of Britain, where spring sowing of cereals is still more common, do some stubbles remain through the winter.

Autumn ploughing covers over much potential food for seed-eaters, and there is a consequent change in the species composition of the bird fauna of fields. This switch has been suggested as an explanation for the reduction in the numbers of some seed-eating species and especially of their wintering populations[1]. Big flocks of seed-eaters are now unusual in southern England in the winter[27]. Moreover there is no renewal of invertebrate food supply in the spring, which will also make the field less suitable.

Other bird species seem to have been affected by this switch to autumn sowing as well. Nesting Lapwings have been noted above, as has the avoidance of autumn sown cereal fields by birds in the winter. It has been suggested that the decline in Song Thrush numbers since the mid-1970s has been caused in part by the modern paucity of the invertebrate food supply in early spring[1], and the case of the Rook is detailed in Box 6.7 since other factors also enter the story.

A practice which has become more prevalent is to dispense with cultivation of the ground between successive arable crops. The new crop is drilled directly into the residue of the previous one, although some of the latter may be burnt off first. Over the years this so-called 'no till' or 'minimum cultivation' leads to higher invertebrate populations in the soil[34], but the invertebrates are usually less accessible to birds and other predators because the soil profile is not subjected to physical disruption. In Iowa the density of nests (of all species combined) on land which was not tilled was 36 per km^2 which was nine times that on land which was conventionally tilled[35]. There were also more species on the plot of 'no-till' (12 vs 3) despite the area of this plot being nearly three times the size of the conventionally tilled plot.

Burning off the straw seems to have little effect on the invertebrates except at the ground surface, but it decreases the organic matter present and may therefore have longer term detrimental effects, although there are as yet no specific studies of the effects of straw burning on the food for birds. Burning of straw is now being severely restricted for

Box 6.7 The Rook and the timing of cereal sowing

The Rook is an omnivorous species which in farmland eats invertebrates and seeds including spilt grain. It relies heavily on fields for its food sources, and the most demanding periods of the year are the early spring, when it starts to breed (egg-laying from mid-March), and in midsummer, when the young emerge from the nest and first become independent. These are both periods when food may be quite difficult to reach, because of invertebrates being deeper within the soil. In early spring this is caused by low temperatures, while in midsummer the ground is often dry and hard near the surface.

The diagram shows the periods when Rooks feed in arable fields (marked with an asterisk) in relation to the stage of the crop. (It must be noted that they also feed extensively on grass fields, especially permanent grazed pasture.) With spring sowing of cereals there is food available for eight or nine months of the year. Seeds and spilt grain are present from about September to February and invertebrates especially in the early spring and just after the harvest. The only period of potential food shortage is the summer. But the situation changes with autumn sowing. Food is then freely available only for about four months, and seed and grain for only a short part of that. Food becomes available earlier in the summer but this is countered by the relative paucity of it during the winter and spring.

Surveys of the Rook distribution and numbers over Britain show that the change in arable regime has had marked effects. Between 1975 and 1980, Rook numbers declined in East Anglia and in central south and southeast England, the regions where the major switch from spring to autumn sowing had occurred, but had remained more or less stable elsewhere[33].

Source: mainly Feare[32].

	January	February	March	April	May	June	July	August	September	October	November	December
Spring Sowing	Stubble ★	Stubble ★	Plough ★	Germinate ★	Germinate ★	—	—	—	Harvest ★	Stooks ★	Stubble ★	Stubble ★
Autumn sowing	—	—	—	—	—	—	Harvest ★	Plough ★	Germinate ★	Germinate ★	—	—

other reasons, so minimum cultivation may become less prevalent as well.

Therefore it appears that spring sowing is much more beneficial to birds in general than is autumn sowing of cereal crops, and that minimum cultivation may have benefits at least for breeding birds.

Harvesting
The ripening crops are a food source for some birds, including House Sparrows, some finches and buntings. Rooks and Woodpigeons may feed on them too, when the crop has lodged or been knocked over by the birds themselves, and being larger birds these can at times cause some economic damage. The harvest itself affects birds because seeds on the ground (including spilt grain) and surface and soil living invertebrates become more accessible when the crop has been removed. The timing of the

harvest is perhaps the most critical factor. Autumn sown crops ripen a month or two earlier than spring sown ones. This means that potential foods for birds are available at different times during the summer, depending on the crop. Again, the case of the Rook is the best studied one – see Box 6.7.

Other operations

Rolling in spring will destroy any nests that exist in the field at the time. Burning off the stubble will certainly remove much of the food present on the surface. Direct drilling means that the advantages of ploughing are not available. Spraying and fertilizing are dealt with below (p. 72). In many cases, though, the details have not been studied and there is no information on the implications for birds.

One aspect which is especially prevalent in autumn sown crops, and which is of positive benefit, is the practice of 'tramlining', where all passes of a tractor through a crop, for fertilizing and spraying, are along the same tracks. The effect on birds is that the tramlines provide a means of easy entry into otherwise fairly impenetrable vegetation. Growing cereal crops are rarely used by feeding birds, although Blackbird, Tree Sparrow and Yellowhammer may do so (see above Box 6.3), but oilseed rape is quite widely used by birds and casual observations suggest that the tramlines are the only means of gaining access.

Changes in farming practice

An increased range of crops within an area will usually lead to an enhanced diversity of bird species and, often, an increase in the density of birds too. Mixed farms generally hold more birds than either intensive arable or purely livestock enterprises (refs 1, 2, and see ch 2).

A factor in arable areas which has to date received rather little attention is the effects of crop rotations in the fields. This has been found to be important for the winter distribution of Lapwing at least[24]. The birds are especially common on cereal crops which have followed leys, and only slightly less so on cereals and grass fields which have followed oilseed rape. This is related to the abundance of invertebrates in the fields.

There have been longer-term changes in the abundances of two other field types which seem to be much used by birds – oilseed rape and fallow. Oilseed rape is favoured by several invertebrate feeders (see above), and the increase in the area sown has probably helped these species. It is also a potential source of seeds for some of the small seed eaters, such as Linnet[1], especially in July and August just before the harvest. The analysis for Box 6.4 found that Linnets show a strong preference for oilseed rape, though there is no evidence for the birds surviving better as a result, or in any way recovering some of their numbers lost since the mid 1970s. Similarly there is no positive evidence for other species either.

The amount of fallow has been decreasing steadily over the last thirty years, such that in 1987 only 0.75% of the tillage crop area of England and Wales was classed as bare fallow in the annual MAFF statistics. However, fallow is one of the options available under the Set-Aside Scheme, and permanent fallow is occupying 79% of land which was set aside in 1988/89[36]. Therefore fallow might become more common if this scheme continues.

Fallow fields are a good source of seeds and invertebrates and many birds, including finches, Stock Doves and Skylarks, will use them as prime feeding sites, especially in the autumn and winter. Fallows are also used as nesting sites by Lapwings and Stone Curlew. Both of these are declining species (see above), and the RSPB is promoting Wildlife Fallow as a preferred option under Set-Aside primarily for them[37]. See chapter 12 for a further discussion of Set-Aside.

In summary, it seems that rotations increase the numbers of invertebrates available and therefore in general increase the numbers of birds. Two other field types have changed in their abundance in recent years: oilseed rape has increased and fallow has decreased considerably. Both are much used by birds.

6.3 Birds and management of grass and livestock

From ecological and wildlife viewpoints the most interesting types of grassland are usually those classified as rough grazing, i.e. unimproved, of low overall fertility and which are grazed at a moderate to low intensity. Conversely, the least interesting are short-term, heavily fertilized leys on drained ground, consisting of only a single kind of grass and which are mown three or four times a season for silage. As usual, though, not every bird species has the same requirements.

It is especially in the lowlands that conservation action is now required to preserve interesting grassland. In 1984 only 11% of the total grassland area of lowland England and Wales remained as semi-natural and rough grassland, and only about 40% of this as semi-natural pasture. Even some of the latter have experienced degrees of agricultural improvement and therefore lost some of their conservation interest[38]. However, land must be managed in order to remain as grassland. If there is no grazing or other removal of woody vegetation, there will usually be natural succession to scrub and (ultimately) woodland.

The greatest bird conservation concern over grassland is for various nesting waders, in particular Snipe, Redshank, Curlew and, to a lesser extent, Lapwing[9], with the Black-tailed Godwit as well in a few particular sites. The Lapwing is discussed above (p. 56) and has rather different habitat requirements to the others, although the Black-tailed Godwit will select areas with some short grass. The godwit and the other three prefer to place their nests within or next to tusocks of rush or thick grasses. All prefer the wetter fields, in part because such tussocks are more numerous, but also because feeding is easier nearby. Snipe, Black-tailed Godwit and Curlew probe into the soil to find food, especially earthworms, and with their long bills they are capable of probing deeper than any other species. Redshank prefer

to feed at the edges of surface pools. Of the four, Curlew are the most tolerant of drier ground. Nevertheless they are confined to unimproved farmland. Some specific recommendations for managing wet grassland for breeding waders are in Box 6.8.

Drainage and grassland improvement

Intensive grassland management involves regular ploughing up and reseeding and there is often considerable use of fertilizers and pesticides. This has several consequences and makes most improved grasslands, and especially leys, less suitable for birds such as waders. Firstly, improved grassland is more uniform in structure, with few or no tussocks. Secondly, it is usually drier. Thirdly, if it is grazed this is usually more intensive which means greater disturbance (see below). Fourthly, if it is not grazed the sward is often too tall and dense for easy movement by feeding birds, especially by chicks. Lastly, there is often less food available because of the use of pesticides.

There has also been a change in the types of leys grown, and this has had one rather different consequence. In the past the most common leys were a grass and clover mixture, with the clover put in to increase fertility. With inorganic fertilisers now so cheap and easy to apply, most leys consist of a dense uniform single-species stand, typically of a rye grass variety. The consequence of reducing the clover content has been that some birds, notably Woodpigeons and Skylarks, which ate the clover and caused some economic damage in the past, have been forced to find alternative food supplies. Unfortunately for the farmers, these are often nearby oilseed rape or other crops which are actually more valuable still.

Drainage leads to a reduction in the length of time in which the land is not subject to direct agricultural use. This is especially important in the spring and early summer. A drained field becomes available for grazing earlier in the season and, when accompanied by increased fertilization, a higher stocking rate of cattle or

Box 6.8 Management of wet grasslands for breeding waders

For nesting, Snipe, Redshank, Curlew and Black-tailed Godwit prefer quite long grass with tussocks, although the godwit additionally requires adjacent areas of shorter grass. Lapwings prefer much more open grassland for their nests. Hence the management to be carried out will depend partly on what species are present and what species it is wished to encourage.

There are three aspects of management which can be adapted quite easily to benefit these birds: water levels, the amount of grazing or mowing, and predator control. The first two will certainly result in some loss of agricultural efficiency, and the third has implications for conserving other birds. Hence the objectives must be clearly and carefully defined before management begins.

Water levels: the ideal is neither too little water, because of making the soil too dry and impenetrable to probe, nor too wet, because of the danger of flooding nests as well as drowning potential food such as earthworms. The optimum is to maintain the water table at 20–30 cm below the surface throughout the breeding season

Grazing and mowing: cattle are preferred to sheep, mainly because they leave some tussocks and do not graze the field to such a uniform close-cropped sward, although this will depend partly on the stocking density. However, if animals are present while the birds are nesting, cattle do more damage with their trampling.

Ideally, cattle should not be put out until after the nesting season is over. For most waders this means not untily early June, although it can be mid-May if only Lapwings are present. For Lapwing, a higher livestock density is needed in the summer to produce the necessary shorter sward by the following spring. At least 250–300 cow days per hectare is recommended for them.

The figures suggest that, at a stocking rate of about two cows per hectare in the nesting fields, Lapwings will lose about 40% of nests to trampling, about 60% of Snipe nests will be lost and 72% of Redshanks. At double this stocking rate, nest losses increase to 60% of Lapwing, 85% of Snipe and 93% of Redshank (R. Green in ref 1, and ref 39).

Predator control: trees, bushes and fence posts can be removed since these are vantage points for such as Carrion Crows (see ch 4). Bushes and trees are advantageous for many other birds, however (see ch 3).

Some trapping or removal of ground predators may be necessary, but large-scale removal of common predators is not encouraged for other reasons, and in many cases is illegal.

Source: especially Green & Cadbury[10], based on RSPB experience.

sheep is possible. Both have detrimental effects, especially on breeding waders.

Rotations in arable farming are mentioned above. Many traditional rotations include a ley, and several birds seem to benefit from this – see Lapwing in winter and Skylark noted earlier.

In summary, draining and improving grass usually lead to fewer nesting birds present. In particular, breeding waders prefer wetter and lightly grazed unimproved grassland.

Grazing

Most studies of the nesting success of grassland waders have found that trampling by cattle is a major cause of nest failure. Because of the increased use of fertilizer and the fields being drier due to drainage, cattle are now let out earlier into the fields from their winter housing. In the Netherlands the whole breeding season of grassland waders now starts two weeks earlier than it did at the beginning of this century[42]. This is thought to be because the ideal grassland structure (length and density) for waders now occurs earlier, together with earlier onset of trampling by cattle, while earlier mowing of the grass destroys later nests. This explanation was disputed by Shrubb, who maintained that a similar apparent effect in Britain was caused simply by fewer lost nests being replaced if there were cattle present in the field[41]. See Box 6.8 for some suggested figures for stocking rates which will not cause undue damage.

Analyses of BTO nest records, especially for the Lapwing, indicate that cattle are much more damaging to nests than sheep[41], and in the Netherlands yearling cattle are particularly harmful[43]. This seems partly so because they are larger and more clumsy animals and therefore trample nests more easily and often, but also, in Britain, cattle are usually managed more intensively, and often on the more improved grass where the stocking rate can be higher. Over the last 50 years the average stocking rates of cattle have doubled[1]. Also, sheep are more often grazed on rough grass where management is generally less intensive. Yet sheep can cause problems also. Where they are stocked at a high density, and where they are given supplementary feeding in the fields in spring, they will produce a sward unsuitable for any birds to nest on.

In summary, grazing is essential to retain areas of shorter grass, but in fields where birds are nesting the numbers of stock should be kept to a minimum and preferably not put out until after the end of the breeding season.

Mowing

The other major aspect of grassland farming which affects birds is mowing, and over 40% of all ley and permanent grass fields are now mown at least once each year[44]. Mowing a grass field has the same effect as harvesting an arable one in exposing invertebrates on and near the surface of the ground, although there are usually fewer seeds available in a mown grass field than on harvested arable land. Most of the birds which use fields only as feeding sites seem to benefit from mowing. Newly mown fields may contain quite large concentrations of several grassland feeding species, such as Starlings and some thrushes – see above (p. 61).

There has, however, been a major change in recent years in the timing of mowing. This is linked to the switch in grassland management from predominantly hay to predominantly silage, at least in the more intensively farmed areas. In the lowlands in the early 1970s some 85% of the fields which were due for mowing were for hay, with only 15% cut for silage. By 1986 the proportion was 70% silage to 30% hay[44].

Silage is cut a month or more earlier in the year than is hay, and there may be two, three or even four cuts in a season, usually four to six weeks apart. The result is that many bird species are now unable to complete their nesting cycle in the grass before the first cut, and the second mowing may destroy the second attempt as well. This change has been implicated in the decline of several grassland species in western Europe, especially the Corncrake, but also others including Yellow Wagtail and some of the waders. Mowing may kill the young birds directly – it did kill about a quarter of Corncrake chicks in the Uists[45] – and it will inevitably expose them to predators just as it exposes invertebrates. Predatory birds and mammals can therefore find the young birds much more easily. Ideally, there would be no mowing of fields which contained nesting birds until the end of the breeding season, which for waders is probably mid to late June. For the same reason there would ideally be no tractor passes over the fields either. Rolling and harrowing certainly destroy nests, but slurry spreading, manuring, fertilizing and pesticide spraying also have some of the same effects.

It is undesirable and probably now impossible from the farming viewpoint to change back from silage to hay. However one simple way of alleviating some of the problems is to mow fields in a different pattern, and this applies as much to mowing for hay as for silage. Farmers usually mow a field in a spiral from the outside inwards. The result is that any birds present become concentrated into the remaining uncut grass in the middle of the field, and to escape they will have to cross a very open area. The simple remedy for this is to cut a field either from one side to the other or from the middle outwards. Both strategies will allow adult birds and their young a better chance of escape, although eggs or young still in the nest will almost certainly be destroyed whatever method is used.

From a botanical point of view it is preferable not to mow a field until the flowers and other plants have set seed, and many of these are adapted to a fairly late cut.

To summarise, mowing will destroy any nests or young birds which cannot escape, while the change from hay to silage has meant earlier and more frequent cutting so that birds have become more vulnerable. However, mowing a field from the inside outwards (rather than towards the centre) will enable at least the more mobile birds to escape.

Supplementary feeding in the winter

An aspect which does not concern grassland management directly, but is closely tied in with it, is the practice of giving supplementary feed to cattle or sheep in the fields in winter. In drier fields many stock are left in the open all through the winter, and they are very often fed with hay or concentrates in the fields, rather than indoors. Both of these systems provide food for a variety of birds, including Starlings and some of the finches and buntings. Yellowhammers have been found to be more common where cattle are fed than away from these[1]. The practice is only likely to have any damaging effects in areas where waders are encouraged to nest, as it allows a higher stocking rate.

Some particular types of grassland

There are a few particular types of grassland which attract a good deal of attention from conservationists, largely because of their rarity and particularly their botanical interest. Types include chalk grassland, acid heaths, and hay meadows which have been managed continuously in a traditional way sometimes for up to a few hundred years. A great deal of conservation money has been spent on retaining traditional management on the few examples of these last which remain unaffected by agricultural improvement. There is now also an economic incentive to keep them, since hay cut from these can attract a premium price, from horse breeders especially. Management is essential to retain them as grassland, or they too will become scrub or woodland, as has happened on many areas of chalk grassland.

It appears that most of these grassland types do not have any particular interest for birds. One relatively large area of traditionally managed hay meadows was studied in Swaledale in the Yorkshire Dales[46]. The main result was that the hay meadows proved to be of very limited interest to ground nesting meadow birds. If the meadows were grazed in winter they had too little vegetation in spring to attract nesting birds. Later in the season they had too much vegetation to attract feeding birds which need short grass. Another, nearby study area, on the Leyburn Flats, had both a mixed farming regime which had replaced the traditional hay meadows, and some rough and wet pastures. The combination of winter stubbles and spring cereals, which attracted Lapwing and Oystercatcher, and the rough grass beside streams which was ideal for Curlew, Redshank, Snipe, Common Sandpiper and Yellow Wagtail, combined with the residual marshier areas as good feeding sites for several species, produced a dense and diverse community of meadow birds[46].

6.4 Fertilizers and other growth promoters and regulators

Two major kinds of fertilizing are practised:

a with manure or slurry
b with inorganic fertilizers, especially nitrates, phosphates and potash.

The considerable majority are now inorganic, especially in arable areas where livestock are rare or absent, because these compounds are relatively cheap and easy to apply. However, manure and slurry are often spread onto fields in livestock areas even if this is only as a means of disposal of these waste products.

There appear to be few (if any) direct effects of inorganic fertilizers on birds although there have been no specific studies. They certainly affect the birds indirectly, though, via the promotion of plant growth earlier in the season. The consequences of this can be interpreted from information already given, especially that on nesting Lapwings. It also appears that some invertebrate groups, in particular earthworms, are reduced when use of inorganic fertilizers is increased[34]. Clearly, this affects the food supply of some bird species.

The presence of farmyard manure on a grass field was found to be one of the most important factors affecting the winter abundance of Lapwings and some other species on fields[20]. There were more birds when farmyard manure had been applied, and the effect was seen both during the application itself and later on, presumably due to increased invertebrate numbers.

So far as is known growth regulators and other such chemicals have no direct effects upon birds in the fields.

6.5 Pesticides

There has been a great deal of discussion and argument about the use of pesticides and what effects they have on wildlife. This is not the place to go into all of the, often bitter, arguments although many of the problems disappeared with the phasing out of the organochlorines. Some particular aspects are mentioned earlier in this chapter and in some others (see for example p. 68 and p. 31), but two general points need reiterating here. Firstly, all pesticides, whether herbicide, fungicide, insecticide or other, should only be applied according to the regulations and instructions for use. If this is done there should be few, if any, accidents or incidents which affect wildlife. Secondly, the chemicals should be applied only where they are needed. In particular, farmers should avoid spraying into hedge bottoms (see p. 31). Both these points make economic as well as environmental sense. A code of practice for the use of pesticides on farms has recently been published by MAFF[47], and should be consulted by everyone intending to use them on fields.

6.6 Summary and management recommendations

This chapter contains various recommendations for the benefit of birds which will certainly impinge on the business viability of the farm. However practices even part way towards the ideal for birds can have considerable benefits, and they need not necessarily involve major costs or inconvenience. The cumulative effect of 'everyone doing a bit' might be large on a national scale. The continually changing range of subsidies available for different practices will also have effects on what is, or is not, viable in economic terms, and those currently being introduced include some which are mentioned below as being of great benefit to birds.

General

1 Bird numbers are usually higher, and there are more species, in areas with a diversity of crop types and management practices.

2 Autumn-sown cereal fields are very little used by birds at any season. Spring-sown cereal fields are strongly favoured for nesting by some species, including such declining birds as Lapwing and Stone Curlew.

3 Broad-leaved crops, in particular oilseed rape, are used for feeding by several hedgerow species, notably Reed Bunting and Blackbird, and a few pairs may even nest in rape, e.g. Reed Bunting and Dunnock. Field-nesting species seem to avoid it, however. In the winter Woodpigeons and Skylarks may eat the young shoots.

4 The most favoured grass type for nesting is the lightly grazed, wetter, rough, unimproved field. The least favoured by most species which are likely to nest on grassland is drained, intensively managed, single-species ley. Long grass is generally avoided at all seasons, although nesting species require sufficient cover.

5 For feeding, most species prefer shorter grass whether this is grazed or mown

6 The feeding preferences can be explained by access. The favoured crops are all easier to move through at ground level and most have been shown to hold more invertebrates.

7 Stubble and fallow fields, both now considerably reduced in numbers due to changes in farming practices, hold many birds of many species where they still occur.

Arable farming

8 Spring-sowing attracts more birds than autumn-sowing, because:
 a cultivating in spring turns up a new potential food supply at a time when other foods are scarce;
 b it provides more open ground in April, which is preferred by several field-nesting species; and
 c stubble is available throughout the winter, not for only a month or less.

9 Rolling, cultivating and other activities in fields will destroy any nests or young birds present. Therefore such activities should be avoided if possible during the breeding season (early April to mid-June) in fields containing nesting birds.

10 Direct drilling and minimum cultivation may have benefits for some breeding birds, although ploughing benefits several species in winter and spring.

11 Straw-burning reduces the invertebrate food supply at the surface, but probably has little longer-term effect on birds.

12 Tramlining for passes of fertilizer and pesticide sprays may offer a means of access by birds to an otherwise inaccessible food supply under the crop.

Grassland

13 Most birds are more common on grasslands of low fertility as these retain a diversity of structure and plant species.

14 Many bird species prefer wetter patches. They are able to probe into the surface, and the soil invertebrates are usually nearer the surface and therefore more accessible.

15 Grazing or mowing is essential to maintain grassland, but a high stocking density reduces the structural diversity and hence the numbers of birds.

16 Cattle are preferred to sheep outside the breeding season, because cattle usually leave some tussocks uneaten. But during the breeding season cattle cause more direct trampling of nests than sheep. Ideally, all grazing should be kept off nesting fields until at least the eggs have hatched.

17 Mowing fields in the breeding season may kill any young birds present, and such mortality is now more common because silage making, with its earlier cutting period, has increasingly replaced hay making. A second cut for silage may destroy a repeat nesting.

18 Newly mown fields, at whatever time, are very attractive to feeding birds.

19 Mowing from the edge of the field inwards concentrates any birds which are present, so that they cannot escape without crossing very open areas. Hence it is preferable to mow from one side to the other or from the centre of the field outwards.

20 In fields where birds are nesting it is preferable to avoid mowing until mid to late June.

21 Chalk grassland and long term hay meadows have no particular attraction for birds, although they are very important in conservation terms for flowers and some invertebrates.

Fertilizers

22 Fields which have had an application of farmyard manure are more attractive to birds in winter than fields which have not.

23 Inorganic fertilizers have few, if any, direct effects on birds, although the indirect effects of plant growth earlier in the season affect several species in an adverse manner.

Pesticides

24 Pesticides should be used only in accordance with the instructions and regulations. Farmers should ensure that the chemicals are only placed where intended and not allowed to drift into (for example) hedge bottoms.

References

1 O'Connor, R.J. & Shrubb, M. 1986. *Farming and Birds*. University Press, Cambridge.

2 Schläpfer, A. 1988. Populationsökologie der Feldlerche *Alauda arvensis* in der intensiv genutzen Agrarlandschaft. *Orn. Beob.* 85: 309–371.

3 Shrubb, M. & Lack, P.C. 1991. The numbers and distribution of Lapwings *V. vanellus* nesting in England and Wales in 1987. *Bird Study* 38: 20–37.

4 Galbraith, H. 1988. Effects of agriculture on the breeding ecology of Lapwings *Vanellus vanellus*. *J. appl. Ecol.* 25: 487–503.

5 Sharrock, J.T.R. 1976. *The Atlas of Breeding Birds in Britain and Ireland*. T. & A.D. Poyser, Calton.

6 Banbury Ornithological Society, 1985. ABSS. Corn Bunting Survey 1985. Ann. Rep. BOS 1985: 33–34.

7 Williamson, K. 1968. Buntings on a barley farm. *Bird Study* 15: 34–37.

8 Møller, A.P. 1980. [The impact of changes in agricultural use on the fauna of breeding birds: an example from Vendsyssel, North Jutland.] In Danish with English summary. *Dansk Orn. Foren. Tidsskr.* 74: 27–34.

9 Smith, K.W. 1983. The status and distribution of waders breeding on wet lowland grasslands in England and Wales. *Bird Study* 30: 177–192.

10 Green, R.E. & Cadbury, J.C. 1987. Breeding waders of lowland wet grasslands. *RSPB Conserv. Rev.* 1: 10–13.

11 Elliott, G. 1988. Montagu's Harrier conservation. *RSPB Conserv. Rev.* 2: 20–21.

12 Green, R.E 1988. Stone Curlew conservation. *RSPB Conserv. Rev.* 2: 30–33.

13 Spott, C. 1989. *The importance of crops as foraging sites for hedgerow birds*. Unpublished thesis for Diplomarbeit, University of Würzburg.

14 Fuller, R.J. 1984. *The distribution and feeding behaviour of breeding songbirds on cereal farmland at Manydown Farm, Hampshire, in 1984*. Report to the Game Conservancy. BTO Research report no. 16. British Trust for Ornithology, Tring.

15 Cracknell, G.S. 1986. *The effects on songbirds of leaving cereal crop headlands unsprayed*. Report to the Game Conservancy. BTO Research report no. 18. British Trust for Ornithology, Tring.

16 Davis, B.N.K. 1967. Bird feeding preferences among different crops in an area near Huntingdon. *Bird Study* 14: 227–237.

17 Hart, A.D.M. 1987. Tree Sparrow breeding behaviour. *The Boxworth Project report for 1987*: 104–113. Ministry of Agriculture, Fisheries and Food, London.

18 Brough, T. & Bridgman, C.J. 1980. An evaulation of long grass as a bird deterrent on British airfields. *J. appl. Ecol.* 17: 243–253.

19 Southwood, T.R.E. & Cross, D.J. 1969. The ecology of the partridge. III. Breeding success and the abundance of insects in natural habitats. *J. Anim. Ecol.* 38: 497–509.

20 Tucker, G.M. 1989. Farmland birds in winter. *BTO News* no. 162: 4–5.

21 Newton, I. 1972. *Finches*. Collins, London.

22 Milsom, T.P., Holditch, R.S. & Rochard, J.B.A. 1985. Diurnal use of an airfield and adjacent agricultural habitats by Lapwings *Vanellus vanellus*. *J. appl. Ecol.* 22: 313–326.

23 Fuller, R.J. & Youngman, R.E. 1979. The utilisation of farmland by Golden Plovers wintering in southern England. *Bird Study* 26: 37–46.

24 Shrubb, M. 1988. The influence of crop rotations and field size on a wintering Lapwing *V. vanellus* population in an area of mixed farmland in West Sussex. *Bird Study* 35: 123–131.

25 Fuller, R.J. & Lloyd, D. 1981. The distribution and habitats of wintering Golden Plovers in Britain, 1977–1978. *Bird Study* 28: 169–185.

26 Green, R.E. 1978. Factors affecting the diet of farmland Skylarks. *J. Anim. Ecol.* 47: 913–928.

27 Lack, P.C. 1986. *The Atlas of Wintering Birds in Britain and Ireland*. T. & A.D. Poyser, Calton.

28 Owen, M., Atkinson-Wills, G.L. & Salmon, D.G. 1986. *Wildfowl in Great Britain*. 2nd Edition. University Press, Cambridge.

29 Department of Agriculture and Fisheries for Scotland 1982. *Wild Geese and Scottish agriculture*. DAFS advisory leaflet.

30 Thomas, G.J. & Owen, M. 1982. *Wildfowl and agriculture*. Advisory leaflet, Royal Society for the Protection of Birds, Sandy.

31 Prater, A.J. 1987. The changing fortunes of the Brent Goose. *RSPB Conserv. Rev.* 1: 47–50.

32 Feare, C.J. 1978. The ecology of damage by Rooks (*Corvus frugilegus*). *Ann. Appl. Ecol.* 88: 329–334.

33 Sage, B.L. & Whittington, P.A. 1985. The 1980 sample survey of rookeries. *Bird Study* 32: 77–81.

34 Edwards, C.A. 1984. Changes in agricultural practice and their impact on soil organisms. Pp. 56–65 in *Agriculture and the Environment* (ed. D. Jenkins). Institute of Terrestrial Ecology, Cambridge.

35 Basore, N.S., Best, L.B. & Wooley, J.B. 1986. Bird nesting in Iowa no-tillage and tilled cropland. *J. Wildl. Manage.* 50: 19–28.

36 Ministry of Agriculture, Fisheries and Food. 1989. John MacGregor announces good start for set-aside. *MAFF News Release* 74/89, 22 February 1989.

37 Osborne, P.E. 1989. *Conservation Advice. The Management of Set-Aside Land for Birds: a Practical Guide*. Conservation Management Advisory Service, Royal Society for the Protection of Birds, Sandy.

38 Fuller, R.M. 1987. The changing extent and conservation interest of lowland grasslands in England and Wales: a review of grassland surveys 1930–84. *Biol. Conserv.* 40: 281–300.

39 Beintema, A.J. 1982. Meadow birds in the Netherlands. *Research Institute for Nature Management Annual Report 1981*: 86–93.

40 Green, R.E. 1988. Effects of environmental factors on the timing and success of breeding of Common Snipe *Gallinago gallinago* (Aves: Scolopacidae). *J. appl. Ecol.* 25: 79–93.

41 Shrubb, M. 1990. Effects of agricultural change on nesting Lapwings *V. vanellus* in England and Wales. *Bird Study* 37: 115–127.

42 Beintema, A.J., Beintema-Hietbrink, R.J. & Muskens, G.J.D.M. 1985. A shift in the timing of breeding in meadow birds. *Ardea* 73: 83–89.

43 Beintema, A.J. & Muskens, G.J.D.M. 1987. Nesting success of birds breeding in Dutch agricultural grasslands. *J. appl. Ecol.* 24: 743–758.

44 *Survey of fertilizer practice*. Fertilizer use on farm crops in England and Wales. Ministry of Agriculture, Fisheries and Food, London.

45 Stowe, T.J. & Hudson, A.V. 1988. Corncrake studies in the Western Isles. *RSPB Conserv. Rev.* 2: 38–42.

46 Fotherby, D.I. 1987. *Birds and hay meadows. Habitat usage in a changing landscape*. Unpublished M.Sc. thesis, University of York.

47 Ministry of Agriculture, Fisheries and Food/ Health and Safety Commission. 1990. *Pesticides: Code of Practice for the Safe Use of Pesticides on Farms and Holdings*. HMSO, London.

CHAPTER 7

Orchards

7.1   Importance of orchards

Orchards and soft fruit production provide habitats for birds that are very different to field crops, and provide different potential resources. A field of soft fruit is similar in some respects to scrub, although scrub has a much more complex structure and therefore higher plant and invertebrate diversities. Similarly, an orchard resembles woodland but it too has a simpler structure. There is usually little if any shrub layer, while the trees are usually densely and uniformly spaced and typically only 5–6 m tall.

No Common Birds Census plots have been predominantly of soft fruit. However, a few consist wholly or predominantly of orchards. Two of these in Kent were censused for several years from 1978. Box 7.1 gives some figures for the birds occurring on the plots.

Orchards support much larger bird populations than do arable crops or grasslands, and thrushes and finches were especially numerous in the orchards censused by the CBC. Almost all of the species which occurred in these orchards nest principally in trees although a few nest on the ground. Birds more typical of shrubs and bushes, including those often in hedgerows, were conspicuously absent from the list of species found breeding within the orchards. For example, Wren, Dunnock and several warblers were all recorded elsewhere on the farms but none were in the orchard trees.

In the breeding season, thrushes and many finches eat mainly invertebrates. However, outside the breeding season, both groups turn more to fruit, or in some cases to buds. Thrushes, including the winter visiting Fieldfare and Redwing, are particularly attracted to fallen fruit in orchards, but they sometimes eat them from the trees as well. Finches feed largely in the trees and when they are eating buds they can cause some economic damage, especially the Bullfinch. Some information on means of alleviating such damage is outlined in Box 7.2.

Box 7.1 Birds in Kentish orchards

Plot A comprised 30 ha, with all of the 'field' area consisting of orchard. Plot B was 71 ha, about half of which was orchard with the rest being arable, mainly spring-sown cereals. On both plots there were hedges of various sizes and a group of farm buildings.

The table shows the total numbers of bird territories recorded on the plots, and the proportion of these which was wholly within the orchard, for the 20 species with at least 10 territories on either plot over the 4 years.

On Plot A 13 species were recorded holding territory wholly within the orchards (those in the table plus Great Spotted Woodpecker), and 11 species were recorded on Plot B. The 'All species' category in the table excludes Woodpigeon, Starling, House Sparrow, Swift and House Martin, because they were either not censused at all or only estimated from a nest count. The Skylark was also excluded because it was impossible to score accurately, especially on Plot B where many territories spanned both orchard and adjacent arable crop areas.

The overall species composition of the two plots was generally similar. The commonest species on both plots were Blackbird, Song Thrush and various finches. In addition, Blue Tits and Turtle Doves were widespread on plot A and Grey Partridge on plot B. The main difference between the orchards of the two plots was that the trees on Plot A were old enough to have some holes present which enabled Great Spotted Woodpecker, Starling and Blue Tit to find nest sites.

Nests of Great Spotted Woodpecker, Blackbird, Song Thrush, Starling, Tree Sparrow, Chaffinch, Goldfinch, Linnet and Lesser Redpoll were found within the orchard on one or other of the two plots.

	Plot A		Plot B	
	Total no. in 4 yrs	% within orchard	Total no. in 4 yrs	% within orchard
Mallard	10	0	2	0
Grey Partridge	0	—	25	72
Moorhen	14	0	0	—
Turtle Dove	31	19	7	14
Wren	48	0	11	0
Dunnock	51	0	32	0
Robin	21	0	3	0
Blackbird	160	21	124	11
Song Thrush	74	20	42	12
Blackcap	13	0	0	—
Spotted Flycatcher	10	10	3	0
Blue Tit	35	20	8	0
Great Tit	21	5	2	0
Tree Sparrow	30	27	14	14
Chaffinch	55	15	40	10
Greenfinch	63	14	45	27
Goldfinch	16	44	8	38
Linnet	24	50	46	35
Lesser Redpoll	19	63	19	47
Yellowhammer	10	0	14	14
All species	777	16	477	19

Source: 4 years of data from 2 Common Birds Census plots.

Box 7.2 Bullfinches and orchards: alleviating potential problems while retaining environmental benefits

At times Bullfinches cause considerable economic damage to fruit trees by eating the buds. However, the extent and timing of damage is unpredictable both on a spatial and a temporal scale. It appears that usually only a few individual birds cause the damage[2].

Several means of reducing damage have been proposed, and have been used with varying degrees of success and impact on other birds and the environment. Control methods have included trapping, shooting, provision of alternative sources of food, spraying repellents and altering the habitat to make it less attractive to Bullfinches.

Trapping and shooting are costly, not least because they usually have to be sustained. Providing alternative food sources has the disadvantage that it may maintain the birds during a period of food shortage, their survival then leading to greater problems later[4]. Repellents have several limitations[5], but may be effective during a period when buds are being damaged[3].

Habitat change is a longer term solution. Experiments on feeding behaviour have suggested that Bullfinches feed predominantly at the edge of orchards and that they prefer not to cross open ground. Therefore creating a barrier of open ground at the edge of the crop may reduce damage in vulnerable areas, or may spread damage more evenly[3]. From an environmental viewpoint it is best to create this barrier by removing orchard trees rather than surrounding habitats as the latter hold more of other wildlife. Furthermore, non-crop habitats may well provide sufficient natural food for the pest species, reducing their need to come into the orchard at all.

Sources: mainly Greig-Smith and Wilson[2], Greig-Smith[3].

A larger scale study in Switzerland[1] involved assessment of all 2599 orchards in the Zurich canton. Most of the areas were small and poorly maintained, and had intensive arable cultivation under the trees. Overall, more birds occurred in those orchards which **a** had many trees with a long trunk, **b** had intensive cultivation under the trees, **c** contained patches of both open and dense vegetation and **d** had other habitat features present (hedges, gardens etc). The Chaffinch was found in 74% of all of them, and Blackbird, Tree Sparrow, Starling, Great Tit and Greenfinch were the next most widespread. There are both similarities to and differences from the two orchards described in Box 7.1. In both areas, tree-nesting birds and species which feed from tree trunks (e.g. Treecreepers) and on fallen seeds dominated the feeding community. However, the Swiss areas held more hole-nesters, probably because the trees were older and therefore contained more suitable potential nest sites.

Another problem for birds is that most orchards are heavily sprayed with pesticides.

Apple orchards are subjected to a greater intensity of spraying, especially of insecticides, than almost any other crop. The effects of this are suggested by the pattern of recovery of the Sparrowhawk population in Sussex after the general low levels in the 1960s. Numbers took much longer to recover in East Sussex than in West Sussex, and this was attributed by Shrubb[6] to drift of insecticides (especially organochlorines) from intensively sprayed orchards in the east. Approval for the use of these chemicals in orchards was withdrawn at a later date than for use on cereals.

7.2 Management of orchards for birds

It is clear from the above that the major reason for the relative scarcity of birds in orchards, compared to woods, is the lack of a shrub layer. However planting shrubs under the trees is not a practical solution for a fruit farmer to remain efficient. Very few of the birds which do occur in orchards nest on the ground, but many feed there. Therefore it is preferable to keep the ground layer open, with relatively short vegetation. Fallen fruit, especially if partly rotten, is a favoured food for many species. Hence leaving some on the ground may prevent birds from pecking at more valuable fruit still on the trees.

7.3 Summary and Management Recommendations

1 Orchards usually hold many more bird species than other crops. Thrushes and finches are likely to be most common and, in the older orchards, hole-nesting species such as tits. Birds typical of bushes and shrubs are almost absent.

2 Fallen fruit may be especially attractive to some birds in the autumn. Leaving some on the ground may reduce bird damage to fruit still on the trees.

3 Bullfinches can cause problems through damaging buds in the spring. Leaving a clear area between the orchard and surrounding hedges can alleviate some of these.

4 Keeping the ground layer vegetation short and open will encourage birds in to feed on the ground.

References

1 Müller, W., Hess, R. & Nievergelt, B. 1988. Die Obstgarten und ihre Vogelwelt im Kanton Zurich. *Orn. Beob.* 85: 123–157.

2 Greig-Smith, P.W. & Wilson, G.M. 1984. Patterns of activity and habitat use by a population of bullfinches (*Pyrrhula pyrrhula*) in relation to bud-feeding in orchards. *J. appl. Ecol.* 21: 401–422.

3 Greig-Smith, P.W. 1987. Bud-feeding by bullfinches: methods for spreading damage evenly within orchards. *J. appl. Ecol.* 24: 49–62.

4 Newton, I. 1968. Bullfinches and fruit buds. Pp. 199–209 in *The Problems of Birds as Pests* (eds R.K. Murton & E.N. Wright). Academic Press, London.

5 Wright, E.N. 1980. Chemical repellents – a review. Pp. 164–172 in *Bird Problems in Agriculture* (eds E.N. Wright, I.R. Inglis & C.J. Feare). Monograph no. 23, British Crop Protection Council, Croydon.

6 Shrubb, M. 1985. Breeding Sparrowhawks *Accipiter nisus* and organochlorine pesticides in Sussex and Kent. *Bird Study* 32: 155–163.

CHAPTER 8

Farm Woods and Scrub

Trees are probably the most important single habitat feature for determining both the overall density of birds and number of species within an area of farmland, and this applies whether the trees are in woods or hedgerows. In this chapter the term 'wood' is used as a shorthand for any group of trees and associated shrubs except for those which are field boundaries of a single line. Few farms include any large area of woodland, but many have small copses or coverts, planted-up corners or shelter-belts. The chapter also includes scrub, which is important for some birds, and which may arise as regenerating woodland, through permitted invasion of bushes or simply by leaving marginal land uncleared. Some of the points in the chapter apply equally to woods and scrub, and a few extra points specific to the latter are considered at the end.

In recent years the most profitable type of woodland in a commercial sense at most sites has been conifers. However, the situation is changing and there are several recent initiatives for promoting more broad-leaved woodland in farmland. The Farm Woodland Scheme[1] and Set-Aside[2] are examples.

Woods are especially valuable for conservation because they hold a relatively high density of animals and plants compared to other habitats on farmland. In addition some of these species are relatively rare nationally and hence of special value. Woods may also be rather more lasting features of the landscape than other habitat types.

Some of the guiding principles for conservation management in woods are listed in Box 8.1. These were drawn up mainly for larger woods, where most of the surveys and studies have been done, but they can be used as guidelines for smaller woods too, although not all the suggestions are applicable directly to farm woods. The latter tend to be smaller and may be fairly isolated. Despite this, some of even the smallest patches have been woodland for up to several hundred years and may contain a few special plants or animals.

All woods which occur on a farm and which could be managed by the farmer are included here, although specialist contractors may be used for larger operations. Larger scale forestry operations are not considered in any detail.

A fair number of farm woods are largely or completely neglected at present, and certainly are not managed for conservation or for any other objective. This may be the best policy, especially for long-established woods (see principles 5 and 13 in Box 8.1), but the majority of farm woods were planted for a particular purpose such as a game covert. If they are to remain useful for conservation, or perhaps to produce a small income from timber production, then active management is often necessary, albeit low key.

The first part of the chapter discusses the importance for birds of farmland woods and what determines how many and which species occur. Management can play a crucial role in this as it can adapt a wood for specific purposes. It is important to note that management of woods is often a rather longer term operation than management of other habitat features in farmland, for the effects may not become visible for several years. The second part of the chapter considers the management options more explicitly, and how they can affect the birds.

With most habitat features in this book, creation of new examples is considered only as an afterthought to managing those which already exist. However, woodland offers more possibilities. This is partly because planting

Box 8.1 Some principles of woodland management for conservation

Peterken's 15 principles for the integration of conservation management with other objectives are primarily directed at larger woods, but many apply to isolated farm woods of any size. The principles are:

1 distinguish between sites of high conservation value (HCV) and others of lower value, both individually and over the surrounding area;

2 set management priorities, perhaps with particular sites having special treatment;

3 minimise clear felling, especially on HCV sites, but avoid filling all gaps;

5 develop or retain blocks of woodland and maintain a scatter of small woods between larger blocks;

6 minimise change, but if change is needed then do it slowly, especially with work affecting landscapes;

7 encourage long rotations or at least some old trees;

8 plant native tree species if at all possible;

9 encourage diversity of structure, species and habitat as far as this is compatible with other aims;

10 encourage restocking by natural regeneration or coppice growth;

11 if necessary, take special measures for rare wildlife species;

12 maintain management records;

13 manage some woods (at least) using non-intervention methods;

14 maintain or restore traditional management where possible and appropriate; and

15 introduce modern management to enhance conservation value if necessary and where traditional management is impossible.

Source: Peterken[3,4].

---

may be an integral part of managing existing woods, but also because there is a variety of grants available, although the exact details of eligibility may change with time.

8.1 The importance of farmland woods for birds

Woodland usually supports a higher density of birds, more species per unit area, and a different range of species than does the complex of fields and hedges, although the size and structure of the wood influence this considerably. About 60% of bird species breeding inland in Britain will nest in woodland[5], with perhaps 10–15% being more or less confined to it for nesting. Additional species will use woods for feeding and shelter at times. Some figures for the numbers of individuals present are given in Box 8.2. These show that the woodland share of the total number of breeding pairs on a farm is disproportionately high by a factor of four or five. For example, when woodland occupies about 5% of the farm area it is likely to hold 20–25% of the breeding birds on the farm, although the exact figure will depend on the structure of the wood and the quality of other habitat features available nearby.

Woods are also of disproportionately high value compared to their area for some other groups of wildlife. Ancient semi-natural woodland (see below) supports more than half the British butterflies and nearly 20% of flowering plants[5].

Box 8.2 The proportion of farmland birds which occur in the woods

Two studies of individual estates provide figures for the proportion of breeding birds which were in woods. Firstly, on the Bovingdon Hall Estate in Essex (total 700 ha) the 11% which was woodland, much of it in one block, was estimated to hold just over 30% of the numbers of breeding bird pairs of fifty species[6]. Secondly, on part (564 ha) of the Manydown Estate in Hampshire about 11.5% was woodland in several blocks of various sizes. These were estimated to hold just over 50% of the breeding passerines[7].

The diagram shows some figures compiled from CBC plots by Hudson[8]. The percentage of the numbers of twenty-six passerine species which breed in the woody vegetation (i.e. not field species such as Skylark) is plotted against the percentage of the area of the plot which was woodland and scrub. Woodland clearly holds a disproportionately high percentage of the breeding birds on a plot.

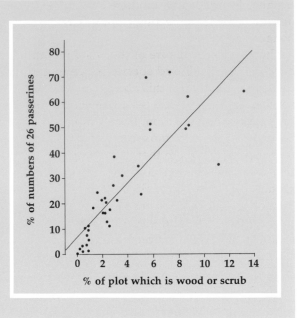

Sources: Mason & Long[6], Fuller[7], Hudson[8].

Some bird species are more dependent on woodland than others. None of the common farmland breeding bird species (Box 1.1) is confined to woods but several are scarce or rare away from them, including Turtle Dove, Garden Warbler, Treecreeper and Bullfinch. Many others are much more common in woods than elsewhere. However, a detailed analysis of the importance of woods for individual species has not been done, nor have there been any studies of the use and importance of woods by birds outside the breeding season.

The most comprehensive study on what affects the numbers of birds and species in farm woods is that by Paul Opdam and some of his colleagues in the Netherlands[9,10]. They studied a series of woods of different sizes, different distances from other woods (which may be sources of immigrant birds), and differing habitat structure. Others have also investigated this. Moore & Hooper[11] visited many woods for short periods and reviewed the British literature, Ford[12] carried out a fairly detailed study of the birds in twenty woods in Oxfordshire, and Shaw[13] studied a series of plots in lowland Scotland. Some general points emerge.

The size of the wood almost always explains a substantial amount of the variation in the number of bird species present, with larger woods holding more species. However, other factors may also have significant effects, for some individual species at least. For example, species which are more or less confined to woodland, not really being farmland birds at all, are less frequent in woods which are distant from others. For some of these birds, this is likely to be because they are reluctant to cross large tracts of open ground. It is noticeable that the effect of distance is reduced when there is a higher density of such connecting elements as hedgerows[10].

The physical structure of the wood has an effect on nearly all bird species which might occur there, and both this and the plant species composition are partly dependent on the geographical location. For example, Scottish woods have more birch and many more conifers in them than English ones and therefore contain more conifer specialist birds such as Goldcrest, Coal Tit and the crossbills. Similarly, Welsh woods are often dominated by the sessile oak, and Pied Flycatcher, Redstart and Wood Warbler are much more common in this type than in others. Overall, there are more breeding bird species in woods in the south and east of Britain than in the west and especially north, and in winter in the southeast and the west than elsewhere[14].

There appears to be a lower density of small birds in areas where the soil productivity is less[15]. Drainage and pH of the soil will also have their effects, although there are no specific data.

Figures illustrating these features are presented in the following paragraphs.

Area of woods

Large woods hold more species of birds than small woods. Box 8.3 gives some specific figures and indicates that the area can explain up to 80% of all variation in the number of species

---

Box 8.3 The number of bird species and number of individuals in woods of different sizes

The first diagram shows the number of bird species found in a wood in relation to the size of that wood. In this study, size explained over three-quarters (81%) of all the variation in the number of resident breeding species. Other studies have shown a similar dominance, with half to three quarters of the variation in bird species number explained solely by the area of the wood (e.g. refs 9, 10, 11, 13).

The second diagram plots the average density of birds against the size of the wood, also from Ford's[12] data. Clearly, density is less in larger woods.

Source: mainly Ford[12].

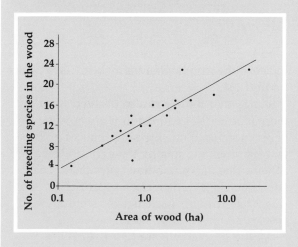

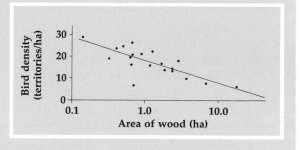

Box 8.4 The occurrence of some individual bird species in woods of different sizes

Three studies have provided some figures:
Oxfordshire[12]: nine species occurred in nearly all the study woods, which varied in size from 0.1 to 18 ha. These were Woodpigeon, Great Tit, Blue Tit, Wren, Blackbird, Robin, Blackcap, Dunnock and Chaffinch.

A further eleven species appeared only in larger woods, although the minimum threshold size differed between species. These were Cuckoo, Great Spotted Woodpecker, Carrion Crow, Jay, Coal Tit, Marsh Tit, Nuthatch, Treecreeper, Song Thrush, Chiffchaff and Goldcrest.

The remaining seventeen species which were found breeding in any of the woods showed no obvious pattern relative to area, although there were more of them in larger woods than in smaller ones.

Lowland Scotland[13]: for eleven bird species, the woods where they were present were significantly larger than woods where they were absent. The species were Pheasant, Wren, Robin, Blackbird, Song Thrush, Willow Warbler, Goldcrest, Spotted Flycatcher, Great Tit, Treecreeper and Yellowhammer.

Britain[11]: In woods of 0.01–0.1 ha, Woodpigeon, Dunnock, Blackbird and

Chaffinch were the only species which occurred, and even these species were often absent.

In woods of 0.1–1.0 ha, the above four species, plus Wren, Robin, Song Thrush and House Sparrow were present in at least 30%.

In woods over 1.0 ha, at least half held Woodpigeon, Wren, Blackbird, Song Thrush and Chaffinch.

Sources: Ford[12], Shaw[13] and Moore & Hooper[11].

---

which occurs[12]. In most cases it has been found that, in order to double the number of species present, the area needs to be increased by a factor of about ten[11].

A large wood will also hold more individuals, but average density is usually less than in a smaller wood. Box 8.3 gives some figures for this too. The probable reason is that small woods have a larger proportion of edge habitat. Edges of woods often have a different habitat structure with, in particular, more shrub and field layer vegetation, both of which

increase the numbers of many birds (see below and ref 16).

Some other figures can be derived from these studies. In Oxfordshire[12] there was never more than one breeding pair of a species in woods smaller than about 0.5 ha, and the average number of pairs per species only reached 2.0 when the area of the wood was 2 ha or more. Some figures for individual species are given in Box 8.4.

All the species which occur in smaller woods are also common and widespread species in other habitats, including hedgerows. It is the less common species, and in particular those which depend on mature woodland, which tend to occur only in larger woods. Woodpeckers, Sparrowhawk, Wood Warbler, Redstart and Woodcock are examples. Further, these woodland specialists are always much less predictable in their occurrence in any particular wood, and there is usually a threshold of size below which they do not occur.

A final point concerns the relative merits of one large wood versus several small ones, and there has been considerable discussion of this in respect of the design of nature reserves. In the context of farm woods, Ford's[12] results showed that a series of smaller woods held more individuals and very often more species overall than one large wood of the same total area, and this was especially so if the small woods differed in their habitat structure. However, if all the woods were small, then some of the scarcer and more specialist woodland bird species would not occur at all. Similar conclusions came from a comparative study of woods in Britain and the eastern United States[17]. The Game Conservancy Trust advocates an ideal, particularly for Pheasant shooting, of several small coverts of 1–5 ha surrounding a central larger wood[16], and this would satisfy the needs for both sizes by other birds as well.

In summary, larger woods hold more species but a lower overall density of birds than do small woods. Many bird species characteristic of mature woods occur only in larger ones, although each species has its own minimum size threshold. Hence a mixture of woodland sizes in a landscape will maximise the numbers of individuals and species of birds that are present.

Isolation

In the context of farm woods, isolation – the distance to the nearest wood – rarely accounts for more than about 10% of the variation in number of bird species, and this factor did not figure significantly at all in one large study[9].

However, as with size, isolation may be quite important for a few woodland specialists which do not readily move from one wood to another. For example Green Woodpecker, Long-tailed Tit and Marsh Tit were found to be less likely to occur in woods which were farther from another wood[10].

A wood within farmland may become effectively less isolated by connecting it to other woods with hedgerows or other non-open habitat. In the Netherlands the number of forest interior bird species was found to be higher when there was a greater density of connecting elements (in this case, wooded banks and rows of trees), but no significant effects on the numbers of any individual species were found[10]. Such connections may be used simply as corridors or stepping stones which enable dispersal to take place while avoiding the need for crossing large areas of open ground, but they may also serve as additional feeding habitat.

Similarly in Scotland[13] six species (Wren, Robin, Willow Warbler, Blue Tit, Coal Tit and Great Tit) were more common in woods where there were more surrounding hedges or lines of trees, and two species (Dunnock and Yellowhammer) were commoner where there was less treeline. Moreover, nine species (Pheasant, Goldcrest, Willow Warbler, Spotted Flycatcher, Redstart, Blackbird, Song Thrush, Great Tit and Treecreeper) were more common where there was less deciduous or mixed woodland nearby, although the first two of these were commoner with more conifer wood present. Only the Wren was more common when there was more deciduous woodland nearby. These figures suggest that there is both a concentrating effect of a wood where there are few other woods present, and an effect of hedges allowing easier dispersal.

In summary, isolation seems to be relatively unimportant for the more common and widespread species, which are characterized by a higher willingness to disperse and wander. But species associated with the interior of woods, and which are perhaps more reluctant to cross open ground, are

usually less common and less likely to occur in isolated woods.

Vegetation structure

The physical structure of the vegetation in a wood is extremely important in determining which and how many birds will occur. There are three major components: vertical structure, horizontal patchiness, and the presence of particular elements such as dead trees. The edge of a wood is a component of the horizontal patchiness but, as it is often rather different to the remainder, it is dealt with in a separate section. More bird species are likely to occur where there is a greater variety of structure and different features.

Vertical structure
Depending in part on the dominant tree species, a mature wood will have three main layers of vegetation: tree canopy, shrub layer and field (or ground) layer.

The main use of the tree canopy by birds is for feeding, as by tits and some warblers. Relatively few species in Britain nest there, although the Rook and Carrion Crow do and the Goldcrest does so in conifers. However, trees are essential to the majority of hole nesters, including the tits, Nuthatch and woodpeckers, in order to provide sufficiently large boughs and trunks for their nest-holes. All these species have been found to be scarcer in woods with few large standard trees[19,20], and in particular the number of natural holes seems to limit the numbers of some of these species in woods, e.g. Pied Flycatcher[21], although this can be remedied at least partially by the provision of nestboxes. Box 8.5 gives some general information about the use of nestboxes.

The extent of the shrub layer is the most important factor for the presence of many bird species. Most warblers and thrushes place their nests there and most insectivorous species feed predominantly within it. A dense shrub layer provides cover for birds and their nests, and shelter which is especially important in winter. Woods with little or no shrub layer have many fewer birds present[22]. However, a more open shrub layer is preferred by some of the woodland specialists, especially those which feed extensively from the ground, for example Redstart, Pied Flycatcher and Tree Pipit. These species are especially common in western oakwoods, partly because these woods have little or no shrub layer and often only a low, grazed field layer[23].

The vegetation at ground level may constitute an important feeding site for several species (e.g. Dunnock and Wren), and a tall dense field layer is used by many of the same species which also use the shrub layer. Bare ground is used as a feeding site by some. The leaf litter harbours a food source for species as diverse as Blackbird, Chaffinch and Great Tit.

In summary, all three layers are important for some birds, especially the shrub layer, but the development of both shrub and field layer vegetation is influenced heavily by the openness of the tree canopy.

Horizontal patchiness
For birds and other wildlife, probably the most important variation across a wood is caused by openings in an otherwise closed canopy. These can occur naturally, as with permanent glades caused by different soil conditions or temporary gaps caused by fallen trees, or by management design, such as creating rides or managing blocks using different harvesting regimes or timetables (see below). The important factor is that opening the canopy lets in light and this promotes growth in the shrub or ground layers. As outlined above, the shrub layer holds more nesting birds than the other layers. Therefore more shrubs will mean more birds. In conservation terms, the Nightingale (in southern England) and some of the warblers are perhaps the most important beneficiaries. Very little is known about the use and significance of such open areas to birds in winter.

The only published study specifically on breeding bird distribution with respect to woodland rides found little effect, but the rides in this case were sharply defined, rather narrow, and not of a structure known to attract birds[16]. To be effective in increasing numbers of

Box 8.5 Nestboxes

Provision of nestboxes is a very easy way to increase the number of potential nest sites for certain species. The commonest type of box is designed for small hole-nesting birds such as tits and the Nuthatch, but boxes can be made for several other common woodland species including owls and woodpeckers. Even Robins, Spotted Flycatchers and some other open nesting species will use artificial sites.

Detailed instructions for the construction and siting of nestboxes for the various species can be found in the BTO Guide[24]. Suffice to say that the common hole type is normally made of wood, and they can be made from offcuts or whatever is available, nailed together to create a waterproof box and fitted with a hinged lid to allow inspection. Size is not particularly critical although obviously they must be large enough. Boxes are best placed on the lee side of the tree trunk or main branch, and away from the edge of the wood. They can be at any height, although if the area is subject to human disturbance it is best to site boxes above the reach of humans on the ground and out of sight of main paths. They can be placed on any trees or bushes including those in gardens, against house walls and in hedges.

In mature or derelict woods, there are likely to be plenty of potential natural sites even for hole nesters. Therefore provision

of nestboxes is unlikely to add greatly to the numbers or diversity of birds which occur. However, in areas with only young trees, where holes are limited, boxes can have a marked effect. This also applies to heavily managed woods with little or no large timber or dead wood present.

Nestboxes also have an educational role. Many people see their first young birds in a nestbox and this can stimulate a longer term interest. People also become very possessive about the contents and follow avidly the fate of the eggs and young.

Source: mainly the BTO Nestbox Guide[24].

nesting birds, rides and glades really need to be wide enough to have a well-developed shrub layer along the side[19]. For this it will need to be at least as wide, and preferably twice as wide, as the height of the surrounding trees. Chiffchaffs have been found to be more common along wide rides than elsewhere in woods, and Willow Warblers can be associated strongly with such rides in woods which lack a shrub layer – in conifer plantations for example[25]. Moreover, a wider ride will develop an open grassy centre, which will be a feeding area for thrushes and some finches, while any shrubs and bushes at the sides will provide nest sites and food sources (e.g. berries) for these and other species.

In older woods a grassy area may well be of botanical interest too as it is unlikely to have been improved by fertilizers. As a result it will often attract insects and invertebrates of conservation interest[26]. Rides are especially important in conifer-dominated woods, where they may provide one of the few areas of broad-leaved vegetation.

Nevertheless narrow rides are sometimes used. Sparrowhawks may hunt along them, Woodcocks may use them for roding, and Tree Pipits can place their territories at junctions of them[26].

In designing rides in woodland it is best to avoid long straight lines, and very desirable to include a bend close to the edge of the wood. The latter will lessen potential wind blow on the trees and will also mean more shelter for any birds and animals in or near the ride and wood[26].

Those glades which are a mosaic of shrubs and open field layer vegetation are usually the most attractive to birds. Glades smaller than about 0.25 ha are much less useful or interesting and for the same reason as narrow rides – less shrubby vegetation.

An area of woodland which is managed actively will usually be more patchy than a neglected wood, because different compartments (panels) will be at different stages of growth. Different bird species are commonest in trees of different ages, so a mixture and diversity of growth stages will lead to a more diverse bird fauna.

Other features which contribute to the patchiness of a wood are sections with different tree species composition (see below), and the presence of ponds and streams. The latter also act to open the canopy, and hence encourage a more lush growth of bushes, while the additional attraction of water can make the area especially attractive to birds. See chapters 9 and 10 for more specific points relating to ponds and streams, much of which applies equally to water in woods as to that in more open situations.

In summary, openings in an otherwise closed tree canopy are especially useful for increasing numbers and diversity of birds, whether the openings are rides, glades, ponds or streams. This is because the increased light allows increased growth of shrubs. Rides are the most easily managed. In general they should be at least 1.5 to 2 times as wide as the height of the surrounding trees and contain a variety of field and shrub layer vegetation.

Edges

It is commonly found that the numbers of animals (including birds) are higher along a meeting of two separate habitats than in either habitat on its own. Birds along woodland edges are no exception, and a third of the bird species occurring in woods are judged to be edge species[19]. External edges, which are those alongside other habitats, are the most important. Internal edges, which are those between different-aged stands or along rides and glades, are much less so. Some figures for birds along woodland/farmland edges are presented in Box 8.6. Clearly, most wood edges provide excellent habitat for birds, although the species composition is not the same as in the interior of the wood.

The main reason is probably the differences in vegetation structure[16]. There can be a considerably greater development of the shrub layer at the woodland edge, and the bird species noted in Box 8.6 all prefer to nest in shrubs or in thick vegetation near the ground. A subsidiary reason is that birds can feed more easily in the adjacent open farmland. This is important for the Pheasant[28], and both the Blackbird and Song Thrush also do this. A further reason for such an effect was suggested in Finland, where the numbers of several invertebrate groups were found to be highest at the forest edges[30].

Some edges seem to be more preferred than others (see Box 8.6). In addition to providing for the habitat requirements of several birds, a hedge around the outside of a wood will provide more shelter near the ground. Pheasants certainly prefer this kind of situation[28]. The addition of a few metres width of shrubs, or even spruce trees, between the edge and the trees will increase the shelter effects and reduce any shading of the crops. It

Box 8.6 The effects of woodland edges on birds

The diagram shows the numbers of records of Blue Tit, Chaffinch, Blackcap and Willow Warbler at different distances from the edge of woodland for 3 types of wood. Similar effects were shown by Wren, Blackbird, Song Thrush and Garden Warbler. Clearly, many more birds were recorded along and near the edge than in the interior of the woods, and no species avoided the edge. The diagram also shows there were much stronger preferences for the edge in unthinned coppice than in either thinned coppice or high forest[16].

A similar preference for the wood edge is shown by the Pheasant[28]. Male birds spent two-thirds of their time within 20 m of the edge of a wood, and only rarely were they more than 50 m in. Pheasants too like edges with a good growth of shrubs, but avoid edges where the trees abut directly onto open fields without a hedge or other shrub layer[28].

In a different study, 19 species of birds used an ungrazed lynchet edge of ash and scrub while only five species used an edge of a young plantation of beech, spruce and cedar[29].

Figure is redrawn from parts of Figs 2 and 3 of Fuller and Whittington[16].

Source: mainly Fuller & Whittington[16].

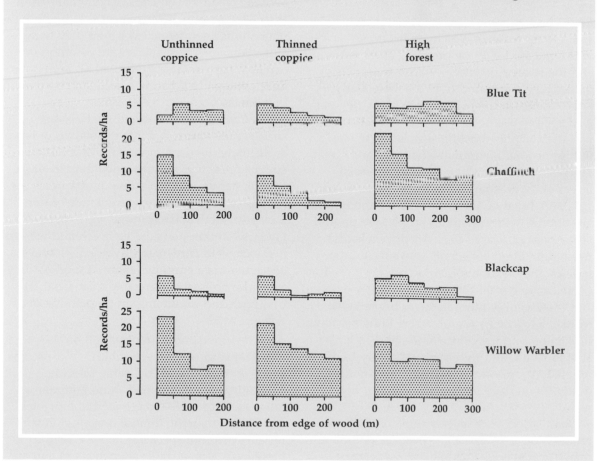

will also create an area from which Pheasants will rise, which is important when a good shooting wood or covert is required[18].

The importance of the edges has considerable implications for the shape of woods. Straight edges will be shorter than irregular ones and, therefore, hold fewer birds, while irregular edges are usually more pleasing from the landscape viewpoint. The edge effect also helps to explain the higher density of birds in smaller woods (see Box 8.3). Small or narrow woods (less than 50 m or so across) can almost be considered as entirely edge.

In summary, the edge of a wood usually holds more birds of more species (especially warblers and thrushes) than its interior. This is due largely to the increased volume of the shrub layer which normally occurs along the edge.

Particular features

The special features which are the most important for birds are probably holes in trees and the related presence of dead wood. Over-mature and moribund trees are more likely to have cavities suitable for use by birds. Several bird species use holes as nest sites and others will use them as roosting sites in the winter, as will some small mammals. Blue Tits and Great Tits need fairly small holes, Starlings need larger ones, Jackdaws and Stock Doves larger still, and Tawny Owls even larger, although the last species will sometimes nest on the ground at the base of a large tree. Woodpeckers also need quite large holes but, unlike most other species, they excavate their own.

Holes are easiest to excavate in softer wood and there is only room in larger branches or trunks. Favoured sites are where branches have broken off and the centre of the broken limb has started to rot, and in dead standing limbs. Great Spotted Woodpeckers rely very heavily for their nest sites on dead limbs and trees and heart rot[27].

Dead wood, especially if some of the bark is retained on it, is also a good source of food for many bird species including woodpeckers.

Great Spotted Woodpeckers get two-thirds of their winter food, November to February, from dead wood and most of that is from dead limbs on live trees[27].

Tree and shrub species composition

The dominant tree species is the main determinant of the structure of the wood. The commonest dominant species in farm woods are probably oak, ash, cherry, birch, elm, alder and sycamore, although several others are widespread and common and may dominate in particular woods or areas. Most farm woods contain a mixture of species. Over much of England conifers are uncommon in farm woods unless they have been planted specifically, but there are plenty of such conifer woods in parts of Scotland.

In general, larger bird populations occur in broad-leaved woods than in conifers, mainly because broad-leaved trees usually have more associated insects, but also probably because plantations are often felled before they mature. However, a few bird species are common in conifers, such as Goldcrest, Coal Tit and Redpoll, and the Chaffinch occurs commonly in both kinds.

Some results from a recent study of tree species preferences by birds are shown in Box 8.7. Clearly there are more birds and more bird species present in mixed woods than in single species stands, and individual bird species can have preferences for particular kinds of tree. European larch and sycamore were especially favoured in this study, while beech and western hemlock were particularly avoided. However beech mast is a strongly favoured food of several species (including Great Tit, Coal Tit, Chaffinch and Brambling) in the autumn and winter.

There is no specific information on the importance or otherwise for birds of particular shrub species, except that some provide berries in the autumn, and those with dense foliage will provide more potential nest sites and shelter than the thinner ones. Again, a mixture of shrub species is likely to lead to more birds than a single species stand, but there is no information. See also the section on shrub species in hedges (p. 23), for much of that applies here as well.

In woods snowberry and some species of honeysuckle are often recommended for attracting Pheasants[18]. Snowberry is an alien species and can become rampant, which is undesirable in a more natural wood being managed for conservation. Both snowberry and honeysuckle provide berries but there is nothing special about them compared with other berry-bearing bushes except that neither is attacked by deer or rabbits. This can have considerable advantages when trying to establish some cover and shelter quickly and cheaply, since fences or guards are unnecessary.

8.2 Management

The management practices to be undertaken in a wood will depend on many factors. These include its current state, whether or not any income is required, the available manpower and the inclinations of the owner. It should be noted too that the requirements of conservation and timber production are not necessarily incompatible.

Traditionally, most woods in Britain were managed actively, often by coppicing, to produce wood of varying grade and quality – see, for examples, reviews and descriptions of the history, present state and management practices in British woods by Rackham[32,33] and Peterken[4]. However, many of the traditional practices have fallen into disuse over the last 50–100 years, largely because markets have declined while the costs, especially of manpower, have risen. The result has been that many woods, especially small farm woods, have become more or less derelict from a management viewpoint, and some fairly drastic treatment would be required to restore them to a state where they could produce a renewable supply of wood. Most such woodlands have a few large trees which are reaching overmaturity (in the forestry sense) and with parts beginning to rot. The shrub layer and field layer are often

completely overgrown by such as bramble, old man's beard and bracken. The shrubs are often few and far between. And there is no chance of an income from any timber which may be present because of its low quality. Such woodlands are often used as a dumping ground for all kinds of rubbish, from hedge cuttings to old cars. Yet, as will be clear from the earlier section, all this is not necessarily bad for birds and some aspects are very good, such as the presence of large old partially rotting timber.

The main management systems

From the conservation viewpoint, the greatest care should be taken with Ancient Woodland. Ancient Woodland sites are those which have borne woodland of one kind or another since at least 1600, and most areas of it in Britain are now registered as such by the Nature Conservancy Council[34]. However, the need for care does not mean that active management should be avoided. Indeed, many of these sites should be managed actively to maintain their interest, whether by coppicing or by managing as High Forest or Wood Pasture.

At some stage over the last thousand years or so many ancient woods have been managed by coppicing although in the majority of cases it has been abandoned in modern times. An actively coppiced wood contains compartments (panels) of various sizes, often very small, which are cut to ground level every 10–20 years. The period depends on the tree species, the commonest being hazel, sweet chestnut and ash. After cutting, the bases of the trees sprout with new stems and after a few years often form a dense impenetrable thicket. The base 'stools' may be very old and very large.

Most coppices also contain some 'standards'. These are trees which are left uncut over several coppice cycles and which form a higher but open canopy above the coppice. In due course these standards are harvested as timber, but not all at once. The commonest standards are oak or ash.

A different management type is Wood Pasture. In this the landscape consists of trees with grazing pasture beneath and very little (if any) shrub layer. The trees may be isolated from each other or form a nearly closed canopy. Also they are often pollarded, which is equivalent to coppicing but at about 2–3 m. This system is now very rare in lowland England although the New Forest retains some examples. However, many of the sessile oak woods of Wales and western England are essentially Wood Pasture in structure because they are open to sheep and deer. In addition, most parkland can be regarded as a form of it.

However, the majority of the smaller farm woods will probably have been managed on the High Forest principle, even if the original purpose was to provide a shelterbelt or a game covert. It has been estimated that 80% of private woods of less than 10 ha were originally planted as, or are only retained as, game coverts[18]. High Forest woods are managed to produce timber. A few originate from natural regeneration or from a singled coppice (a coppice from which only one stem has been allowed to grow on from each stool to near maturity), but most are from planting following a clear fell. The continuing management usually involves felling individuals or groups of trees as required and then replanting the gaps.

Active management

One of the most important points to remember with woodland management is that, from a purely conservation point of view, there is often no urgency to do anything. In many woods, nature has taken its course, and will continue to provide a quite acceptable habitat for many birds. Often it is when other factors arise, such as a need to extract some wood or the desire to increase the shooting potential, that active

Box 8.7 Preferences by birds for different tree species

The first diagram (redrawn from Fig. 4 of Peck[31]) shows the number of bird species recorded breeding in a forest compartment in relation to the number of tree species present. Clearly, the number of bird species was higher with more tree variety, and even the addition of a few individual trees of a different kind can add bird species. The overall density of birds was also higher with more tree diversity.

The second diagram illustrates the preferences between March and October of six bird species for feeding in six tree species common in farm woods. The 6 species are BT=Blue Tit, GT=Great Tit, CT=Coal Tit, GC=Goldcrest, CH=Chaffinch and TC=Treecreeper, with the 'All' column being these six species combined. The birds concerned will have been eating invertebrates at the time. Sycamore was favoured strongly and beech avoided, with the other tree species being preferred by some birds and ignored by others.

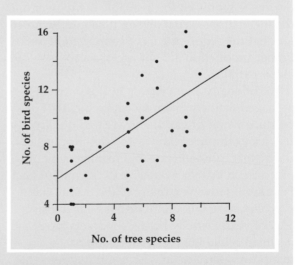

Outside the breeding season some of these birds turn more to alternative foods. In particular, beech mast is a strongly favoured food by many species (including Great Tit and Chaffinch) in years when there has been a good crop.

Source: mainly Peck[31].

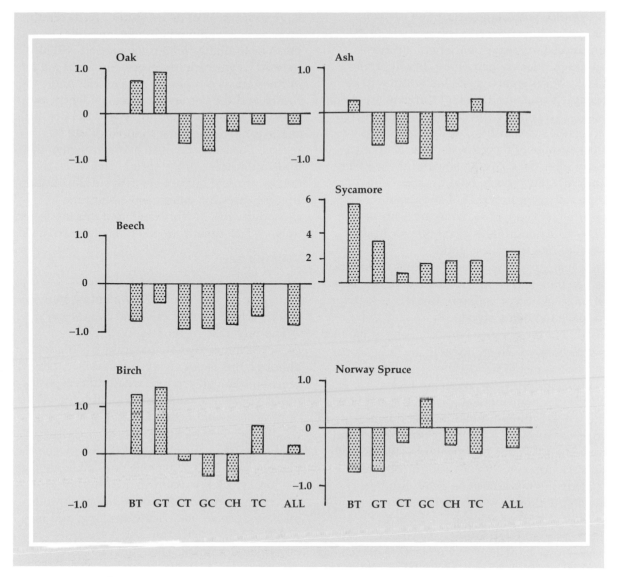

management is begun.

Earlier sections of this chapter described the important factors in woods for birds, and any management undertaken should bear these in mind. The simplest, and almost the only way to manage the smallest woods, is as High Forest. This can usually be achieved best by thinning out and replanting. Thinning can cause a rapid increase in bird numbers because it encourages the shrub layer, and eventually it will lead to taller trees. The oldest stands of High Forest have been found to hold the highest overall bird densities[20,35]. Old trees, an open canopy and a dense shrub layer are the most important features of these. In such situations any new trees planted can be allowed to grow up with little or no management required, except perhaps for protecting young trees from fast-growing weeds and grazing animals (see below).

Unless there is a market for such as fuel wood, and usually only if the wood is more than 4–5 ha in size, it is unlikely to be economical to renovate or start up a coppicing system. In a smaller wood it is impossible to coppice a sufficient area even every third year or so for the full range of stages (the ideal for birds) to be present.

Coppice-with-standards, however, is promoted heavily for nature reserves as it produces the greatest variety of structures, although it lacks dead wood. It will promote blocks of different age stands which will be occupied by a succession of different species of birds. In a coppice system the highest overall diversity and density usually occurs at the thicket stage of the growth cycle, at about 4–10 years after the cut, with some of the warblers and, in a few woods, Nightingales being especially common then. Landscape requirements advocate compartments of 0.25–0.5 ha and this is also good for birds[36]. Smaller blocks are less good.

When coppicing, some standards should be left as they will be needed by any bird species of more mature woodland. But it is essential not to leave too many standards for they will then shade out the coppice beneath. Forty standards per hectare, of which up to 15 are large, is the maximum recommended[36].

Problems with the coppicing system are that it is very labour intensive and the market for the wood is often limited. In addition, neglected coppice stools may have lost their vigour and may be vulnerable to browsing, especially by deer. Hence, before renovating a coppice, consideration ought to be given either to leaving it alone – which obviously is no good if wood or an income is required – or to conversion to High Forest. The latter can be done either by felling and replanting, which has considerable landscape implications, or by singling although this is only suitable for certain species such as ash and lime. In larger woods, felling in blocks but leaving shelter belts or isolated trees – a system which is quite common in continental Europe – is preferable, from both bird and landscape viewpoints, to clearing trees completely. For example, overall bird density in such shelterwood systems, and some of the species dependent on mature trees (such as the Nuthatch) are retained, until the shelterbelt itself is felled by which time the regeneration of the felled area is proceeding well[37].

Many woods, even semi-derelict ones, will have some areas of dense shrubs. Birds tend to be especially common in these, and piles of brash or thinnings left on the ground will also be used. Ivy, which has been considered a bane of foresters, is a useful cover for nests and roosts, and the berries are eaten avidly by many birds in the late winter (ref. 38 and see Box 3.8).

The timing of management work can be critical. Particular times to avoid are when the birds are nesting in the spring, and when the berries are ripe but not yet eaten in the autumn. In most cases the winter is the best time to carry out work in woodland, and this tends also to be a slack time from other farming activities.

Planting trees

An important aspect of most active management in woods is tree planting both in woods which already exist and in new ones.

The preferred method for birds and conservation is to allow natural regeneration, although this is not always practicable. Natural regeneration will produce a natural composition of native species, whilst individual trees will be of local stock and hence adapted to local conditions. So even in plantations it is desirable to allow some natural regeneration between the planted trees. Of course, this does presume that the local stock is suitable in quality.

Natural regeneration can be rather slow, it may produce an inadequate stocking, and may result in one or more unwanted species becoming dominant. Therefore managers often resort to some planting. Following the principles of natural regeneration, native species and local stock should be used whenever possible, and a range of tree species is likely to be more interesting than a single species stand. It is important not to forget to plant some shrubs as well as trees.

The various combination of species, mixtures and planting densities have few effects on numbers and species of birds other than those noted above. Details of suggested mixtures and densities for different regions and soil types are given by several organisations, including the Forestry Commission[1,39], the RSPB[40] and MAFF[41].

In some situations young trees are potentially vulnerable to weeds and grazing. Mechanical weeding is preferable to the use of herbicides, although if the latter are used carefully (e.g. spot treatment with propyzamide or glyphosate) they will not cause any lasting harm unless the existing ground flora is interesting in itself. Plastic sheets on the ground around individual trees are surprisingly effective although messy. Young trees should not need fertilizers but, as with herbicides, a little used carefully is unlikely to do any serious harm unless (again) the other flora is adapted to very particular conditions.

On large plots, protection from grazing and browsing is best achieved by fencing the whole area, although this can be very expensive. Deer and rabbits need a high fence and a small mesh respectively to keep them out. However, a small mesh size will also hinder the movements of Pheasants and other ground birds. Individual tree guards may be more economical for small plots (less than about 1 ha) and there are various types available.

A common practice, and one which has several advantages for conservation and economics, is to plant a 'nurse' crop in amongst the longer-term trees. Very often this will be a conifer or other fast growing species. They will serve as protection and shelter from extreme conditions for young slower-growing trees, as well as for any birds and other animals present. When these are later thinned out, they may serve as a crop in their own right.

Siting new woods

A new wood is likely to be more interesting to wildlife, and be colonised more quickly and easily by animals and plants, if it is associated with other habitat features. In particular, planting more trees adjacent to an existing wood, or incorporating a pond or hedge, is preferable to planting trees in the middle of a field or other open area. Sites which have been wooded in the past may still contain a remnant seed bank which will flourish. However, wood edges which are already of special interest should be avoided as sites for new plantings.

The other major factor to consider when planting is the impacts on the landscape. It is important to remember it is not just the initial planting but the subsequent management which may have marked effects.

8.3 Field corners and isolated trees

In a landscape otherwise devoid of trees, a small group planted in a field corner may act as an attraction to birds. A field corner planting is essentially just a very small wood and most of the points noted above will apply. When there are more trees in the vicinity, a small group of trees or bushes at such a corner is likely to serve as a focus for the territories of some birds. Birds anyway concentrate at hedge intersections[42] and any extra woody vegetation at such sites will enhance the effect.

Groups of trees may also be planted as a screen for buildings or other structures, or as shelter for livestock. These will have effects on the local bird populations similar to those of other groups of trees in a comparable position (see above and ch 11).

Single trees might possibly act as stepping stones for a few birds, but any trees which are a long way from others are unlikely to be a major attraction for the birds of the area. They can, however, have considerable value as landscape features.

8.4 Scrub

Importance and use by birds

Scrub is often thought of as a poor substitute for woodland, but really it ought to be considered as a habitat in its own right. Woodland and scrub have many bird species in common, but scrub can provide a habitat for species which are absent from woods that are lacking in young vegetation. An area of scrub will provide nest sites for several warblers, finches and Turtle Doves, and the density of warblers can be very high. As most scrub areas consist largely of berry-bearing bushes, they also attract thrushes in the autumn and early winter, and these species seem to prefer the greater shelter and protection afforded by areas of scrub to those of hedges.

As with woodland, a structural variety will attract more birds of more species. Scrub which has foliage down to the ground holds many more, especially warblers[14]. Old and 'leggy' scrub which is open underneath holds few breeding birds, like similar hedges. Again as for woods, openings such as rides and glades can further enhance the numbers of birds present. For example, on Manor Farm in Wiltshire, a 1 ha block of dense and relatively uniform scrub, surrounded by heavily grazed grassland, contained four breeding pairs of four species. A nearby block of the same size but with some open patches, and with a longer edge surrounded by rough downland, contained 21 pairs of 13 species. The openings also attract more Fieldfares, Redwings and finches in the winter[43].

Management

If left unmanaged, scrub will become denser, the canopy will close and eventually most areas will become closed woodland. Each successional stage holds a slightly different community of birds, and the denser scrub will usually have a higher bird density as long as it does not become too 'leggy' (see above). Therefore, to retain a diversity of birds, management should be adjusted to retain several of the stages of scrub succession from fairly open grassland to quite dense thicket[14,44].

Scrub development has been seen as a problem in a few areas, notably on some chalk grasslands (downlands) which have a rich and unusual flora. In the past such areas also held some of our rarer breeding birds, including the Stone Curlew and Cirl Bunting, though the latter species requires some bushes and trees nearby as well. Both are now rare species in Britain. In such special situations, conservation in a wider context, although not usually for birds, may be served best by clearing away all the woody vegetation in an attempt to restore the earlier ground flora. This is most readily achieved where the scrub canopy has not closed and some grassland remains. However, if the canopy has closed or the scrub is not growing on land with any particular conservation value the value of the original flora is likely to have been lost. Then, more species and individuals of birds will use the land if it is managed as scrub with, perhaps, some parts allowed to develop into woodland.

To keep an area as scrub, the woody vegetation will need to be cut periodically. Depending on the plant species and productivity of the land, rotational cutting should suffice to retain a patchy area suitable for a higher density of birds. Some control of rabbits may also be required, since scrub is a favoured haunt for them. As in other habitats, active management should be avoided in both the nesting and the berry seasons.

8.5 Summary and Management Recommendations

Importance of woods in the landscape

1 Woods usually hold the highest density of birds of all habitat features on farmland, and more species per unit area.

2 The proportion of the breeding birds on a farm which are in the woods is usually 4–5 times the proportion of the area occupied by the woods. For example, woods which occupy 5% of the area will hold 20–25% of the breeding birds.

3 There is little or no information on the importance of woods outside the breeding season, but the situation is likely to be similar because many of the same factors apply then.

Influence of different factors on bird numbers and species

4 The most important factor determining the number of species in a wood is its area. Larger woods support more species, especially those of mature woodland.

5 Smaller woods hold a higher density of breeding birds, probably due to the high proportion of edge habitat available.

6 The isolation of a wood seems to play little part in determining the number of bird species present although there are usually fewer species, particularly of mature woodland specialists, in those woods more distant from other woods.

7 Connecting elements, such as hedges, partially alleviate any effects of isolation.

8 A more diverse vertical structure of the vegetation increases numbers and species of birds. A well-developed shrub layer is especially important and for warblers and thrushes in particular.

9 Rides and glades increase the diversity of vegetation structure, and increase the numbers of birds correspondingly. Rides should be at least as wide as the height of adjacent trees, and preferably more.

10 The edge of the wood is particularly important, and can support especially high densities of some species such as the warblers.

11 Dead wood and holes in trees are vital for several species. Even woodpeckers which excavate their own holes prefer dead branches for this.

12 Broad-leaved trees hold more birds of more species than do conifers, although there are a few specialists in the latter.

13 Some bird species have particular tree species preferences, so that more tree species means more birds (numbers and species).

14 Woods with a surrounding hedge or which are associated with a pond, hold more bird species than those completely isolated.

15 Soil productivity and presence of other bird species have only a small effect on numbers and species composition of birds.

Management

16 Management of woods is often needed to maintain or enhance the numbers and diversity of birds. Most woods will have been managed at some stage in their history.

17 Special care should be taken with Ancient Woods, but active management should be undertaken where and when required.

18 Most small farm woods have been managed on the High Forest system. A continual long-term rotation of felling and planting is usually preferred, because it retains the landscape and structural features throughout the cycle.

19 Thinning can lead to an increase in bird numbers, because of the promotion of growth of the shrub layer when more light can penetrate the top canopy.

20 Coppicing can be an excellent means of increasing bird numbers, but is impractical in small woods, and there may be labour and marketing difficulties.

Planting trees

21 Planting is usually an integral part of management. Natural regeneration is preferred, if possible, as it creates a natural mix of tree densities and locally adapted species. Planting should be primarily of native trees.

22 Young trees may need some protection from weeds and grazing animals. Herbicides (if used with care), plastic sheets or mechanical means are practical for weeding, while fencing (large areas) or tree-guards (small areas) are best for preventing grazing or browsing.

23 Siting new woods near existing ones is likely to ease natural colonisation of animals and plants, but existing edges already possessing a rich flora and fauna should be preserved.

Field corners and scrub

24 Field corner plantings can be considered as very small woods, and all the above points can be adapted accordingly.

25 Scrub has many similarities to woodland but has a distinctive bird community. To retain this, some active management is required to maintain the diversity of structure.

References

1 Insley, H. (ed.) 1988. *Farm Woodland Planning*. Forestry Commission Bulletin no. 80. HMSO, London.

2 MAFF. 1988. *Set-Aside*. Ministry of Agriculture, Fisheries and Food, and Welsh Office Agriculture Department, London.

3 Peterken, G.F. 1977. General management principles for nature conservation in British woodlands. *Forestry* 50: 27–48.

4 Peterken, G.F. 1981. *Woodland Conservation and Management*. Chapman & Hall, London.

5 Nature Conservancy Council. 1977. *Nature Conservation and Agriculture*. NCC, London.

6 Mason, C.F. & Long, S. 1987. Management of lowland broadleaved woodland – Bovingdon Hall, Essex. Pp. 37–42 in *Conservation, Monitoring and Management* (ed. R. Matthews). Countryside Commission, Manchester.

7 Fuller, R.J. 1984. *The distribution and feeding behaviour of breeding songbirds on cereal farmland at Manydown Farm, Hampshire, in 1984*. Report to the Game Conservancy. BTO Research report no. 16. British Trust for Ornithology, Tring.

8 Hudson, R. 1988. *Bird territories in relation to habitat features in CBC farmland data*. Report to the Royal Society for the Protection of Birds. BTO Research report no. 33. British Trust for Ornithology, Tring.

9 Opdam, P., Rijsdijk, G. & Hustings, F. 1985. Bird communities in small woods in an agricultural landscape: effects of area and isolation. *Biol. Conserv.* 34: 333–352.

10 Van Dorp, D. & Opdam, P.F.M. 1987. Effects of patch size, isolation and regional abundance on forest bird communities. *Landscape Ecology* 1: 59–73.

11 Moore, N.W. & Hooper, M.D. 1975. On the number of bird species in British woods. *Biol. Conserv.* 8: 239–250.

12 Ford, H.A. 1987. Bird communities on habitat islands in England. *Bird Study* 34: 205–218.

13 Shaw, P. 1988. *Factors affecting the numbers of breeding birds and vascular plants on lowland farmland*. NCC Chief Scientist Directorate commissioned research report no. 838.

14 Fuller, R.J. 1982. *Bird Habitats in Britain*. T. & A.D. Poyser, Calton.

15 Newton, I., Wyllie, I. & Mearns, R. 1986. Spacing of Sparrowhawks in relation to food supply. *J. Anim. Ecol.* 55: 361–370.

16 Fuller, R.J. & Whittington, P.A. 1987. Breeding bird distribution within Lincolnshire ash-lime woodlands: the influence of rides and the woodland edge. *Acta Oecol.* 8: 259–268.

17 McClellan, C.H., Dobson, A.P., Wilcove, D.S. & Lynch, J.F. 1986. Effects of forest fragmentation in New- and Old-World bird communities: empirical observations and theoretical implications. Pp. 305–313 in *Wildlife 2000* (eds J. Verner, M.L. Morrison & C.J. Ralph). University of Wisconsin Press.

18 McCall, I. 1988. *Woodlands for Pheasants*. Game Conservancy Trust, Fordingbridge.

19 Fuller, R.J. & Warren, M.S. 1991. Conservation management in ancient and modern woodlands: responses of fauna to edges and rotations. Pp. 445–471 in *The Scientific Management of Temperate Communities for Conservation* (eds I.F. Spellerberg, F.B. Goldsmith & M.G. Morris). British Ecological Society Symposium no. 31.

20 Smith, K.W., Averis, B. & Martin, J. 1987. The breeding bird community of oak plantations in the Forest of Dean, southern England. *Acta Oecol.* 8: 209–217.

21 Sharrock, J.T.R. 1976. *The Atlas of Breeding Birds in Britain and Ireland*. T. & A.D. Poyser, Calton.

22 Yapp, W.B. 1962. *Birds and Woods*. University Press, Oxford.

23 Stowe, T.J. 1987. The management of sessile oakwoods for Pied Flycatchers. *RSPB Conserv. Rev.* 1: 78–83.

24 du Feu, C. 1989. *Nestboxes*. Field Guide no. 20. British Trust for Ornithology, Tring.

25 Fuller, R.J. 1991. Effects of woodland edges on songbirds. Pp. 31–34 in *Edge Management in Woodlands* (ed. R. Ferris-Kaan). *For. Comm. Occ. Paper* no. 28.

26 Warren, M.S. & Fuller, R.J. 1990. *Woodland Rides and Glades: their Management for Wildlife*. Nature Conservancy Council, Peterborough.

27 Smith, K.W. 1987. The ecology of the Great Spotted Woodpecker. *RSPB Conserv. Rev.* 1: 74–77.

28 Hill, D. & Robertson, P. 1988. *The Pheasant. Ecology, Management and Conservation.* BSP Professional Books, Oxford.

29 Matthews, R. & Rowe, J. 1987. Management of mixed woodland – Manor Farm, Wiltshire. Pp. 42–46 in *Conservation, Monitoring and Management* (ed. R. Matthews). Countryside Commission, Manchester.

30 Helle, P. & Muona, J. 1985. Invertebrate numbers in edges between clear fellings and mature forests in northern Finland. *Silva Fennica* 19: 281–294.

31 Peck, K.M. 1989. Tree species preferences shown by foraging birds in forest plantations in northern England. *Biol. Conserv.* 48: 41–57.

32 Rackham, O. 1976. *Trees and Woodlands in the British Landscape.* J.M. Dent & Sons, London.

33 Rackham, O. 1986. *The History of the British Countryside.* J.M. Dent & Sons, London.

34 Kirby, K.J., Peterken, G.F., Spencer, J.W. & Walker, G.J. 1984. Inventories of ancient semi-natural woodland. *Focus on Nature Conservation* no. 6. Nature Conservancy Council, Peterborough.

35 Fuller, R.J. & Taylor, G.K. 1983. *Breeding birds and woodland management in some Lincolnshire limewoods.* BTO research report no. 9. British Trust for Ornithology, Tring.

36 Fuller, R.J. & Warren, M.S. 1990. *Coppiced Woodlands: their Management for Wildlife.* Nature Conservancy Council, Peterborough.

37 Smith, K.W. 1988. Breeding bird communities of commercially managed broadleaved plantations. *RSPB Conserv. Rev.* 2: 43–46.

38 Snow, B.K. & Snow, D.W. 1988. *Birds and Berries.* T. & A.D. Poyser, Calton.

39 Hibberd, B.G. 1988. *Farm Woodland Practice.* Forestry Commission Handbook no. 3. HMSO, London.

40 RSPB. No date. *New farm woods and birds.* Conservation Management Advice booklet. RSPB, Sandy.

41 ADAS. 1988. *Practical Work in Farm Woods. 6. New Planting.* Advisory Leaflet P3165. Ministry of Agriculture, Fisheries and Food, Alnwick.

42 Lack, P.C. 1988. Hedge intersections and breeding bird distribution in farmland. *Bird Study* 35: 133–136.

43 Rowe, J. 1987. Scrub management – Manor Farm, Wiltshire. Pp. 75–79 in *Conservation, Monitoring and Management* (ed. R. Matthews). Countryside Commission, Manchester.

44 Fuller, R.J. 1987. *Composition and Structure of Bird Communities in Britain.* Unpublished Ph.D. thesis, University of London.

CHAPTER 9

Ponds and Standing Water

Ponds, and indeed all waterbodies, are attractive features of any farm for many birds and other wildlife. The water itself is responsible for some of this. Moorhen, Coot, Mallard and other ducks, for example, will not be present unless there is a pond or a stream, and in very dry weather any water will attract many other species to drink or bathe. Another very important factor is the location of the pond with respect to other features nearby, and these other features are sometimes the primary attraction to birds. A great many farm ponds are next to a hedge or in a field corner and many farmyards have a pond close by.

This chapter concentrates on small farm ponds. Some comments are appropriate for larger waterbodies too, but on these there are often additional potential activities such as

shooting, angling or even boating. None of these latter are considered here, but shooting and angling at least are not necessarily incompatible with good conversation, although restricting access for these to only part of the bank may be preferable. The requirements for shooting and wildfowl management are described in some detail in a recently published Game Conservancy book by Michael Street[1]. This deals mainly with lakes and larger ponds and should be consulted by those whose interests lie in that direction.

9.1   The importance of ponds

In an analysis of the importance of several habitat features to farmland breeding birds, the

Box 9.1 Birds and the removal and creation of ponds

Plot A – pond drainage. A pond of about 3 ha was drained, filled in and converted to arable land over the winter 1981/82. The number of bird territories in the vicinity was scored for the four previous years (1978–1981) and the four years afterwards (1982–1985). In the diagram the total number of territories in these two groups of years for Mallard (MA), Lapwing (L), Moorhen (MH), Coot (CO) and Reed Bunting (RB), all of which are more or less dependent on water, and for all species combined (ALL) are shown by the solid bars, together with the number expected in the second period if only the national population trends had been operating (the dotted bars). The number of species

involved was twenty three before and seventeen after the pond was drained. Clearly, numbers of all five of these individual species were considerably reduced, and numbers of several others were also down with the removal of the pond.

Plot B – pond creation. In the winter of 1978/79 a new circular pond about 60 m across (0.3–0.4 ha) was dug, and a few new trees were planted, in a field near to a farmyard on a mixed enterprise farm. The number of territories in the area was scored as for Plot A but for three years on either side, i.e. 1976–1978 and 1979–1981. The diagram plots the numbers of territories of the same group of species, again with the expected numbers if only national trends had operated. The number of species recorded was sixteen before and eighteen after the pond was dug. In this case the new pond brought in some aquatic species which had not been present previously, but had relatively little effect on overall numbers of birds in the area.

Source: 2 CBC plots in Yorkshire.

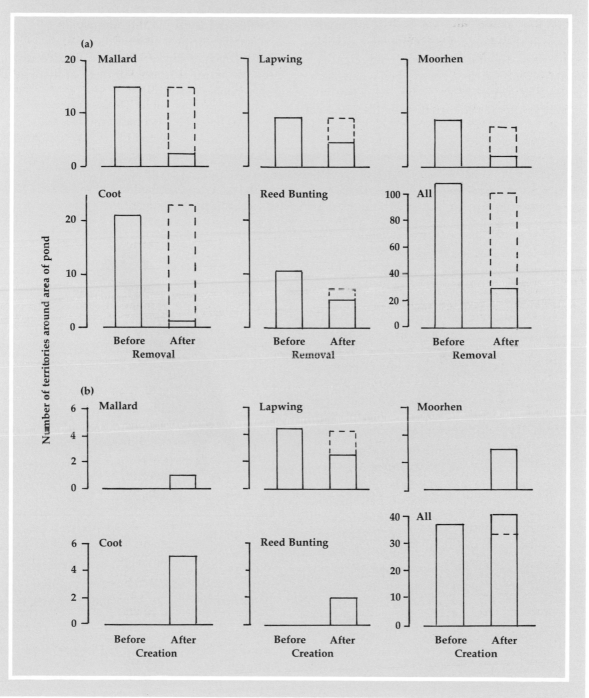

density of thirty three species (of the fifty seven analysed) was associated with the number of ponds on the farm[2]. However, ponds were the most important habitat feature for only five of these species. Hence ponds seem to be a significant general feature in the landscape for many birds but vital for rather few.

Another way of showing their significance is to look at the effects of removing an established pond, or of digging a new one. Some figures from Common Birds Census plots where each of these has occurred are in Box 9.1. Clearly, aquatic birds only occurred when the pond was present, and several other species also benefited even though they did not necessarily use the pond itself directly.

Only two of the common farmland breeding species (Box 1.1) can really be considered aquatic birds – Mallard and Moorhen – although such passerines as Pied Wagtail, Sedge Warbler and Reed Bunting are usually associated with wet areas. Other aquatic species, such as ducks, Coot and Reed Warbler, occur on farms at times, but most of them need larger pools or streams.

Box 9.2 The occurrence of aquatic birds on ponds of different sizes

The first pair of diagrams shows the number of ponds of different sizes studied in Huntingdonshire[3] and over lowland England[4], and the number which had Moorhens breeding on them. Clearly Moorhens occurred on ponds of all sizes although they were more likely to occur on larger ones. Only ten of the Huntingdonshire ponds were not occupied and the reason was clear for seven. They were prone to drying up, were adjacent to occupied ponds or there was no suitable cover for nests (see Box 9.3).

Other species mainly occurred on larger ponds. In Moore & Hooper's[4] study, Coots were found on 80% of ponds larger than 1050m[2], but on none smaller than this, and the Great Crested Grebe was only found on those of more than 1.5 ha (on 50% of ponds of 1–10 ha, and on five of six ponds of more than 10 ha). Other species which occur on larger ponds include Little Grebe, Mute Swan, Canada Goose and other ducks.

The third diagram shows some results for the Coot and Mallard, as well as the Moorhen, from a study in eastern Jutland, Denmark[5]. Clearly both Coot and Mallard

occurred on a much greater proportion of larger ponds than smaller ones, but the Moorhen showed no particular preferences for any size within the range.

Sources: Relton[3], Moore & Hooper[4] and Jorgensen[5].

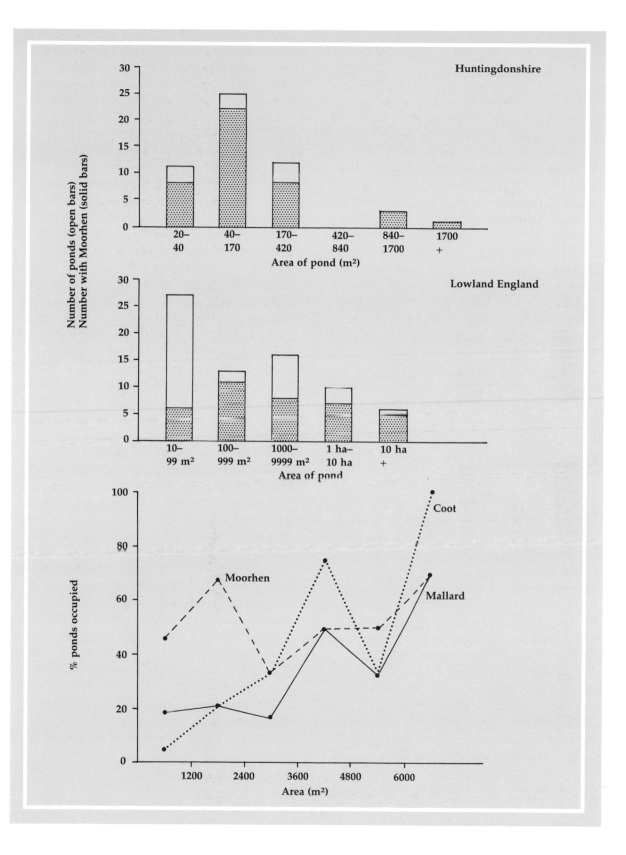

9.2 The influence of physical factors

Size and surface area

Many farm ponds are rather small, only up to about 30 m across or about 0.1 ha, although a few farms have lakes, reservoirs or old mineral workings which are considerably larger than this.

The Moorhen is by far the commonest bird associated directly with ponds. Its only real need for the water is as a safe place to nest, for it feeds mainly on the surrounding land. Mallard too are not totally dependent on water. Females may place their nests some way off, and they too feed partly from the land. Some figures for the use of ponds of different sizes by Moorhen, Mallard and some other species are in Box 9.2.

Others among the aquatic bird species usually occur only on the larger ponds. This is because, unlike Moorhen and Mallard, they mainly feed in the water.

There are many more aquatic birds of all kinds in Britain in winter than in the breeding season[6,7]. Several species of duck enter Britain in quite large numbers and will utilise any open water, even fairly small ponds at times, although those less than about 1 ha are only likely to attract a few Mallard and perhaps the occasional Teal.

Among birds which are not directly dependent on the water itself the size of the pond seems not to very important. Sedge Warbler, Reed Warbler, and Reed Bunting depend on the emergent aquatic vegetation, with the Reed Warbler almost exclusively dependent on the common reed, but the amount of vegetation is not determined only by the size of the pond (see below).

Other species associated with water include the Pied Wagtail, which often feeds along the edge, and Swifts, Swallows and other martins which often congregate over water to feed on the plentiful insect supply. House Martins also need mud for their nests.

In summary big ponds will hold considerably larger numbers and varieties of birds although Moorhens may occur even on the smallest. At least 1 ha seems to be necessary for many ducks and grebes, and this applies both in summer and winter.

Depth of Water

The size of a pond and its depth tend to be closely related but even some small ones, especially those dug for water storage, may be fairly deep. The depth has considerable consequences for the life in a pond, largely because in deeper water there will be much less light and therefore less vegetation, and the temperature will remain more constant through the year. In summer the surface layer, especially of shallower ponds, may become quite warm and too warm for some animals and plants.

The Moorhen and Mallard are largely unaffected by differing depths of water although they can only feed in shallower areas. Both species upend and reach down into the water to feed on any vegetation growing there, but they do not dive. Many of the other aquatic species survive very well in shallower ponds too although diving species such as Coot, Tufted Duck and Little Grebe usually feed 1–3 m down.

There is no obvious advantage for farm ponds to be deeper than about 2 m, as relatively little plant life will grow below this but it is essential to have some shallower areas. Ideally the edge should shelve gently into the water on at least one side, to enable easy access and a variation in depth across a pond will accommodate the various preferences. Shallow areas should be wide enough to support plenty of vegetation (see below) and a strip around the edge of the pond at least 1 m wide is often recommended.

The most important areas of a pond are shallower areas, and these are of most use around the edges. There is no advantage to water deeper than about 2–3 m.

Water quality

The quality of the water in a pond affects birds only via potential food supplies. Thus the most useful ponds for birds and other wildlife contain clear well-oxygenated water with a variety of plant growths. Unfortunately many farm ponds are not like this. Smaller ones are often stagnant, overgrown and almost devoid of life. They may even be used as dumps for farm waste, old cars and other rubbish, none of which is conducive to a healthy pond and some of which may be pollutants.

Stagnation and eutrophication occur when there is too little oxygen in the water to support plants and hence animals. The cause is usually that the oxygen is being used up by micro-organisms breaking down rotting organic matter, and the process may be fuelled additionally by effluent and fertilizer runoff.

When restoring a pond which has been ignored and stagnated, or when siting and digging a new one, the first priority is to prevent runoff of fertilizers and other effluents from getting in and continuing the eutrophication process. It will then be necessary to remove and subsequently keep out such dead vegetation as fallen leaves and dead submerged plants. This should be done physically, as herbicides are normally disastrous to an aquatic environment. In the longer term, cutting back overhanging bushes or tree branches will help to keep out falling leaves, and the introduction of suitable water plants will help recovery and maintenance, although some physical 'cleaning' will probably remain necessary from time to time.

It is essential to keep out potentially poisonous chemicals. Effluent from silage clamps, especially from bad silage, is extremely poisonous, and both this and the effluent from animal houses add considerably to the organic content of a pond and hence eutrophication. Runoffs from surrounding fields may be more difficult to avoid, but maintaining a 'wild' area between the pond and any crop, perhaps by planting with bushes or trees, will certainly help as will directing main field drains away from the pond. Other potential pollutants to avoid are oils and fine suspended matter, as both will lead to less light penetration through the water.

If the pond is very acidic, adding some crushed limestone or chalk can improve the quality of the water and hence the numbers of animals and plants in it, but it must be done carefully and only after careful testing beforehand.

In summary, clear well-oxygenated water is the ideal to aim for. Effluents from silage and animal houses, and runoff of fertilizers and pesticides should be directed away from the pond.

9.3 Vegetation

In the water

The aquatic vegetation in and around ponds is perhaps the most vital factor in promoting their attractiveness for birds. Some plants in the water are essential and the most useful ones are those which are good oxygenators, and those which are eaten by birds or other animals. Duckweeds and pondweeds are good for both purposes, and others include hornwort, water milfoil, and water starwort. Water lilies add colour and their leaves may be used as a means of access by small birds, but the tough stems can cause problems when cleaning, and both water-lilies and some other floating species are liable to become too dominant and shade out other plants.

Apart from those used as food, the plants of most use to birds are emergents. This is mainly as cover and support for them and their nests. The common reed is one of the most familiar plants in lowland ponds and provides excellent cover. Others include bullrush, branched bur reed, and yellow flag. The last is often used as the base for Moorhen or Coot nests although both birds usually use other vegetation to build the nest itself. Moorhens, Coots, Mallards and Little Grebes all hide in emergent vegetation, especially when they have young, while Sedge Warblers and Reed Buntings often feed there even if they nest in the surrounding bushes. Even Reed Warblers may colonise a pond if there is a reed margin at least a metre or so wide. This species is expanding its range

Box 9.3 The amount of emergent vegetation and birds

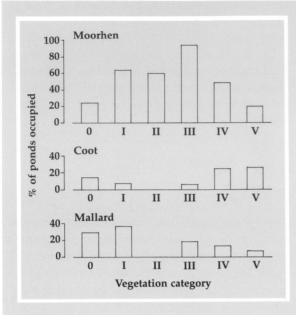

In the Danish study quoted in Box 9.2[5] the amount of emergent vegetation in the pond was also measured. It was scored into one of six categories: 0 – none (twenty one ponds), I –narrow fringe around up to ¾ of the edge (fourteen ponds), II – a patch of vegetation in the middle (five ponds), III – a wide block at one end (seventeen ponds), IV – vegetation all around but with an irregular shaped area of open water in the middle (seventeen ponds), and V – no open water (sixteen ponds).

The diagram shows the proportion of ponds of each type which contained Moorhen, Coot and Mallard. Moorhens clearly preferred those with some vegetation but not those with no open water, Mallard preferred the more open ponds and Coots showed no clear pattern.

Source: Jorgensen[5].

northwards and westwards in Britain although it is still rare in most of Wales and very rare in Scotland.

Reed beds are also a favourite roosting site for several bird species especially in early autumn, and even reed beds only 100 m² or so may attract a few birds. Pied Wagtails, some buntings and, before they leave in the autumn for their winter quarters, Swallows and Yellow Wagtails, all roost regularly in reeds over water. Starlings also prefer to roost there during the autumn but, as they usually roost in groups of several thousand, they are restricted to the larger sites.

One problem with emergent vegetation, including reeds, is that silt accumulates around the base with the result that, over the years, the area covered by the plants steadily dries out, thus reducing the area of open water. If the water area of the pond is to be retained then the reeds or other vegetation must be checked every two or three years. Regular cutting will be sufficient in many cases but more drastic measures, such as dredging and digging out may be required every ten years or more.

How much vegetation to keep in the pond is not an easy question to answer. Box 9.3 gives some information for Moorhen, Coot and Mallard and shows that these species need some vegetation but may avoid those ponds with dense vegetation all across them. Moorhens and other aquatic birds also prefer to be able to swim into vegetation, rather than walk in, probably because they are then safer from land-based predators.

In summary therefore some floating and some emergent vegetation is essential for the retention of many bird species in the pond. The main uses are as food and cover.

In the surrounds

Most ponds have some bushes or trees alongside, rather than being in an open field. But to improve the quality of a pond it is best not to have it completely surrounded by woody vegetation. The southern side is likely to be the best one to leave open. Therefore it will also be the best side on which to have shallower areas of water and sloping banks. Woody vegetation on the other three sides will also give some shelter from the prevailing colder winds.

Several birds are attracted to feed on any soft and damp ground around a pond, especially in dry weather. Song Thrushes in particular use such areas and they are attracted in winter too[8]. Such feeding on the ground obviously requires some fairly open areas on at least some edges of the pond, but most of the birds likely to use them are species of hedgerow or other woody vegetation, and they like to be able to retreat to this if and when they are disturbed by potential predators.

To avoid too many leaves and other vegetation falling in, it is best to avoid having too many branches overhanging the pond. On larger ponds designed to attract wildfowl there

should be a sufficient flight line for the birds to get in and out. Normally ducks will not use a pond where the angle in any direction is steeper than about thirty degrees from the top of the vegetation to the water surface[1]. Therefore it may be necessary to restrict the height of the taller trees.

Some waterside birds will only occur if there is vegetation nearby. For example, larger ponds which have fish may attract Kingfishers. But these will be unable to use the pond for fishing unless there are suitable perches from which to search (see Box 10.3 in the next chapter for specific details of the requirements of Kingfishers).

Woody vegetation around a pond provides some cover and protection from predators and partial protection from the elements. Some open areas, especially with damp ground, are also vital for non-aquatic birds. Hence a variety of vegetation around the sides is a best compromise.

9.4 Other factors

Islands

Islands can be a great attraction to many birds, especially wildfowl, the main reason being that islands are usually free of ground predators. To ensure this safety, the island really needs to be 20 m or so from the shore, so it is only worthwhile creating one in larger pools. There will need to be a sloping bank on at least one side to enable the birds to get onto the island and, for wildfowl to nest, some cover will be necessary in the form of thick grass or preferably low bushes. As long as the island is more than a few square metres, there is the potential for several pairs of wildfowl to nest, notably Mallard and Tufted Duck. Both these species are increasing as nesting species over much of England and both will nest only a few metres apart at times.

Islands can be created by putting in piles of earth or simply leaving them when the pool is being dug in the first place. In the former case it will almost certainly be necessary to shore up the sides with wood (preferably) or other material, at least until it is well established. The alternative is to create a floating island anchored to the bottom. These have been very successful in places, especially in smaller and deeper ponds, and in a few places in southeastern England they have attracted Common Terns and Ringed Plovers to nest.

Introducing animals

A great many animals live in freshwater. Some are aquatic for only part of their life cycle, examples being some insects, frogs and other amphibians, and many of these have a remarkable ability to find newly created and isolated ponds. Other species are entirely aquatic and will be unable to disperse to a new isolated pond.

Some of these animals contribute to making a pond healthier, and therefore should be encouraged. Often the easiest way to introduce a variety of animals to a new pond is simply to empty a few bucketfulls of water from an already existing healthy pond. With a new pond it is also a good idea to move some vegetation taken from an established pond as there will often be invertebrates attached. This kind of introduction is obviously non-selective and it has more potential for ending up with a natural mix of species. Small fish are essential for some of the birds mentioned above, such as Kingfishers, but they should not be put into small ponds because they will eat everything else.

Siting and building a new pond

A potential difficulty when digging a new pond is to have a watertight lining. If necessary a lining of puddled mud, clay or even a butyl sheet may have to be added to the bottom before filling with water. In such cases, it will also be necessary to put some topsoil or other base, such as straw mixed with farmyard manure, onto the lining in order to create suitable conditions for plants to become established.

Without a continual inflow of water, new ponds are liable to dry out, and especially if the

water is pumped out for any reason such as for filling animal troughs. Those ponds with any parts deeper than 2–3 m should retain water even in the driest periods. Note that taking water direct from a stream, or damming and diverting one as a water supply for a new pond, requires permission from the National Rivers Authority.

Care is needed when siting a new pond. It is usually best to site it near to other habitat features of interest to birds and other wildlife which may colonise. Ponds adjacent to hedges, small woods, field corners and other ponds are likely to gain and retain more animals and plants naturally than, for example, one sited in the middle of a field.

9.5 Summary and Management Recommendations

General and Importance of Ponds

1 Ponds and other standing waters are attractive features of farmland to many birds, although many are attracted to the surounding vegetation as much as to the water itself.

2 Ponds are especially attractive in dry weather and, if not iced over, in freezing conditions also.

3 Analyses suggest that ponds are important general features but vital for rather few bird species. Only two common farmland birds are aquatic, Moorhen and Mallard, but others, including Pied Wagtail, Sedge Warbler and Reed Bunting, prefer the vicinity of water.

Physical Features

4 Moorhens will occur on almost any size of pond, as will Mallard, although the latter is rarer on the smallest ones.

5 Larger ponds or pools (larger than 1 ha) may attract Coots, Tufted Ducks, Little Grebes, Canada Geese and even Great Crested Grebes and Mute Swans. The first three all need some vegetation as cover and all do as nest material. All use the water as feeding sites as well.

6 Lakes may attract a variety of wildfowl in winter. Aerial access involving a flight angle no steeper than about 30 degrees over surrounding vegetation should be ensured.

7 A variety of depths creates a useful mixture of conditions. For many birds the shallower parts are of the most use, and there is little or no advantage for ponds to be deeper than about 2–3 m. Shallower areas near the edges are particularly good, and a sloping bank for easy access on at least one edge is essential.

8 Water quality is important indirectly. Effluent from silage clamps and animal houses, in particular, should be directed away from the pond, as should fertilizer and pesticide runoffs. All these will foster eutrophication. Similarly a pond should not become a dump for rubbish.

Vegetation

9 Some aquatic vegetation is essential as food or cover, but dead and rotting vegetation (e.g. fallen leaves) and masses of green algae should be avoided. Physical cutting and removal may be required periodically, perhaps accompanied by dredging or digging out at longer intervals.

10 Herbicides should be avoided if possible, as they are generally very toxic in aquatic environments.

11 Reeds provide excellent cover around pond edges but may have to be cut back periodically to prevent them overrunning the whole pond.

12 Woody vegetation surrounding the pond is beneficial as cover and shelter, but should not be allowed to dominate completely. In particular, it is a good idea to open one side to let in warmth and light, and this will also enable ground-feeding birds to reach the water and any damp ground nearby.

Other Factors

13 Islands are a great attraction to birds, especially wildfowl, because of the protection afforded against potential ground predators. Ideally, the islands will be at least 20 m from the edge of the pond and will have a sloping bank on at least one side.

14 Ideally, new ponds should be sited to link with other existing habitat features. Plants and animals can be introduced as appropriate though some will colonise naturally. It is best to avoid introducing fish until the plants are well established, and it is important to avoid overstocking, especially with predatory animals.

15 Larger ponds and lakes have the potential for other activities, e.g. fishing, shooting for wildfowl and even boating. Fishing and shooting are not necessarily incompatible with conservation for other birds (see ref 1 for shooting) but it is best to keep a part of the water and banks free of disturbance.

References

1 Street, M. 1989. *Ponds and Lakes for Wildfowl*. Game Conservancy, Fordingbridge.

2 O'Connor, R.J. & Shrubb, M. 1986. *Farming and Birds*. University Press, Cambridge.

3 Relton, J. 1972. Breeding biology of Moorhens on Huntingdonshire farm ponds. *Brit. Birds* 65: 248–256.

4 Moore, N.W. & Hooper, M.D. 1975. On the number of bird species in British woods. *Biol. Conserv.* 8: 239–250.

5 Jorgensen, O.H. 1975. [Breeding birds in farmland ponds of Djursland, eastern Jutland, 1973.] (In Danish with English summary.) *Dansk. orn. Foren. Tidsskr.* 69: 103–110.

6 Lack, P.C. 1986. *The Atlas of Wintering Birds in Britain and Ireland*. T. & A.D. Poyser, Calton.

7 Owen, M., Atkinson-Willes, G.L. & Salmon, D.G. 1986. *Wildfowl in Great Britain*. 2nd Edition. University Press, Cambridge.

8 Moles, R. 1975. *Wildlife diversity in relation to farming practice in County Down, N. Ireland*. Unpublished Ph.D. thesis, Queen's University, Belfast.

CHAPTER 10

Rivers and Streams

This chapter is concerned with permanently flowing watercourses. Such a definition includes some ditches around fields, and there is no real sharp division between these. Ditches are considered specifically in ch 5, although that is concerned mainly with those which are not major water channels.

All watercourses serve an essential agricultural function in land drainage. Today, most agricultural land is drained, and the maintenance of many waterways is regulated by the needs of land drainage and flood protection. All maintenance is under the general supervision of the National Rivers Authority (NRA) although some of the work on the banks is the responsibility of the owner, often a farmer. Box 10.1 notes some of the legal aspects and constraints on managing rivers and streams

Box 10.1 Legal aspects of watercourse management

Rivers and streams are subject to many more legal constraints and regulations than other farmland habitat features, because of their wider influence. These include flooding, drinking water, sewage and effluent disposal, and because the same river may flow through areas a hundred kilometres or more away from the farm.

Ultimate responsibility for river management work lies with the National Rivers Authority, although it may contract out the actual work involved. However, much of the work done on the banks is the responsibility of the riparian owner.

In carrying out management work the NRA has a legal duty to further conservation as far as this is compatible with other requirements. There are also some legal constraints on what farmers can and cannot do. For example, a farmer cannot block a river, extract water from it or stock it with fish without permission. The function as a drain must be kept open and maintained, this including removal of any trees or branches which fall in. Moreover, the farmer must not do anything

which would affect the use of the waterway by anyone with legal access to it, and this includes the general public where there is a public right of navigation or right of way along the bank.

Source: various.

which have to be a background to any management for birds or other aspects of conservation.

All waterways, whether they are natural rivers or artificial canals, are important for other reasons. They are often attractive corridors through a landscape with a variety of different vegetation and habitat features associated, some have historical significance, and they are the focus of many and various leisure activities from swimming and boating to fishing. They are also often associated with other wetlands, and these are becoming increasingly rare in lowland Britain with many resources directed towards conserving them (see ch 6). Luckily, for most conservation and landscape purposes the primary considerations are the same. A waterway which looks attractive in a landscape will usually also be interesting and attractive to birds and other wildlife.

10.1 Value of rivers and their use by birds

Of the common farmland breeding birds in Box 1.1, the following are more or less dependent on water: Mallard, Moorhen, Yellow and Pied Wagtails, Sedge Warbler and Reed Bunting. All six will also use ponds and lakes, although streams are probably the more important for Mallard, Yellow Wagtail and Reed Bunting.

In the lowlands some less common species are dependent on water although all will use ponds and lakes as well as rivers and streams. Various ducks, Little Grebe, Coot, Grey Heron, Kingfisher, Sand Martin and Reed Warbler are among them. In the higher reaches of streams, where the water flow is faster, then Common Sandpiper, Grey Wagtail and Dipper are species largely confined to streams. All three occur on Common Birds Census plots but on too few for any worthwhile analyses to be made, although they occur more often on Waterways Bird Survey plots[1].

Most of these species use rivers all the year, while in winter the rivers may become important for others too, at least on a temporary basis. Rivers may be used as a refuge in freezing weather by birds normally more associated with standing water, e.g. Coots, several ducks and Grey Heron. One of the most attractive birds of rivers is the Kingfisher. Box 10.2 describes some of its requirements.

Lowland canals are more or less equivalent to slow flowing rivers, although disturbance is likely to be greater along those which are still used for navigation.

Clearly rivers and streams are important habitat features for several birds although relatively few species are confined to them.

10.2 Physical features of streams and their effects on birds

In terms of physical characteristics much the same applies to streams as to ponds (see p. 104). Shallow areas are more important to birds than deep, a sloping bank allows easy access, and wide rivers will provide more protection from ground predators than narrow. However none of these is under the farmer's control.

One aspect that is under control of the farmer concerns runoff, and it is even more important for streams than for ponds (p. 105) because the effects can extend well down stream. There are, for example, special regulations for the use of pesticides near to watercourses, and farmers have a legal obligation to ensure especially that the two worst potential polluters, farm slurry and silage effluent, do not reach watercourses. Both these are extremely poisonous and can effectively kill quite long stretches of river.

Vegetation in the waterway

Less vegetation will grow in faster flowing stretches of streams, largely because it cannot root and take hold on a rocky substrate. However, even the fastest flowing rivers have hollows, eddies and backwaters where some soil accumulates and plants can become established. Vegetation in the water is essential for many birds either as food (e.g. Mallard and Mute Swan) or as shelter and cover (most

Box 10.2 Encouraging Kingfishers on streams

The Kingfisher is probably the most spectacular of Britain's commoner river birds. It is commonest in the south and in the lowlands largely because other areas only rarely satisfy its particular habitat requirements. It is sensitive to water pollution because of the effects on its food supply and it is very susceptible to severe winter weather.

The Kingfisher's food is mainly small fish, such as minnows and sticklebacks, though it will also eat tadpoles, larger insects and other invertebrates. It gets almost all its food by diving into the water and prefers to search from overhanging perches 1–2 m from the surface which have a clear view underneath, though it will also hover before diving in.

Along a waterway a Kingfisher will occupy a territory all the year round of usually 1–2 km length, and for feeding it will need a succession of suitable perches. However, it is potential breeding sites which are often in limited supply, especially on managed rivers. The essential requirement for a nest site is a vertical bank which the birds can dig into, preferably 1 m or so from the water surface. The birds prefer a bank with water at the base because it gives greater protection from predators, and a bank which is being actively eroded because it is likely to be easier to excavate. Stabilized banks often acquire a tough crust. Kingfishers also like to have a few roots or twigs nearby on which to perch, but sites with a lot of vegetation are avoided, probably due to the holes then being more accessible to ground predators. A common place for a nest site along managed stretches of stream is in a bay or oxbow which has been left after a channel has been straightened.

Managed waterways usually have sloping banks which are unsuitable and some have very few suitable perches for hunting from.

Vertical earth banks are also an essential requirement for nests for the Sand Martin. This species will nest in lower banks than Kingfishers, and usually colonially in more open sites.

Sources: various.

species). The species of plant in slow flowing streams and canals are often the same as in ponds, and serve the same purpose for birds – see ch 9.2.

The banks and immediate surrounds

The banks of a waterway and the immediately adjacent ground are probably more important for conservation and more amenable to management than the water itself. This is partly because the watercourse imposes some constraints on how the surrounding land can be used, and as a result it is often less intensively farmed than sites farther away.

The ground in river valleys is usually rather wetter than elsewhere and may be subject to periodic flooding in the autumn and winter, albeit for less time than in the past because of better drainage. Yet even a short period of flooding may be long enough to restrict the growing of arable crops, so that grass and livestock are predominant in many river valleys. Wetter grass usually holds more breeding birds than other types of fields (ch 6) although outside protected areas there are now only rather few breeding waders[3,4]. Floods may attract many wildfowl and others.

The bank of a watercourse is rarely cultivated so close to the edge that there is no strip of rough grass or other semi-natural vegetation. Sedge Warblers and Reed Buntings are common in most such areas and several other species will use any trees present (see p. 25).

Trees along watercourses are often pollarded. In the past this provided an income from the wood, but pollarding is often still carried out today to prevent the trees from becoming top-heavy and then falling into the stream. For birds, pollarded trees provide more, and often larger, holes than intact trees and several species used them as nesting and roosting sites. Little Owls, Tawny Owls and Tree Sparrows are often recorded in them (CBC data).

By their very nature, semi-natural strips along watercourses will act as corridors for wildlife. The Barn Owl is also known to use them as hunting areas, and the Hawk and Owl Trust has been advocating strips at least 5 m wide along the banks of rivers especially for this species[5].

In summary, the vegetation along the banks of a watercourse is very important indeed for birds. Rough vegetation and pollarded trees provide habitats which are increasingly rare in modern farmland. Because main river valleys are often still subject to periodic flooding, the surrounding land may be less intensively farmed and therefore more attractive to birds.

10.3 Management of Watercourses

The primary object of the management of watercourses is to ensure that their drainage function is maintained as efficiently as possible. In the past this was often done with no thought for conservation, with the result often being that the watercourse changed from being productive and rich in wildlife of all kinds to a uniform bare-sided channel with little or no cover and of very much less interest for wildlife. With the agricultural need for drainage one cannot propose that all vegetation is left to grow unchecked, although this is often the way to promote the greatest variety of wildlife. However, with some careful thought and design, it is possible to leave some vegetation *in situ* and thus reduce the often considerable effects of insensitive management. Two examples of the effects of insensitive management are in Box 10.3, and see also Box 4.5 on the management of ditches.

The following paragraphs discuss some of the management operations which are carried out and suggest ways in which these can be modified to be less destructive to birds and their habitats, and yet which are relatively minor as far as cost and work involved are concerned. It is important to note that most effects of waterway management on birds are through the effects on the vegetation. Much of the work on the waterway itself will not be under the control of the farmer, but it is included here as a guide and because many of the comments are applicable to large ditches as much as to major watercourses.

The first, general, point concerns the timing of any work, especially any involving removal of vegetation. As in the case of hedgerows, the most critical times for birds are the breeding season when they have nests, and the autumn when many of the shrubs have berries. The winter is perhaps the least damaging season in which to carry out management work on waterways.

Box 10.3 The effects of insensitive stream management on bird numbers

Plot A: Moorhens were censused along a small river and a canal as part of the BTO's Waterways Bird Survey[6]. Parts of both waterways were subjected to different levels of management by dredging and removal of bankside vegetation off both sides. The graphs show the number of territories recorded per km on each section of each waterway. The numbers of Moorhens were reduced considerably on recently managed parts of both. Moorhens also bred later in the season, nested less frequently near the channel and produced fewer second clutches on the managed sections.

The main reason for all these differences

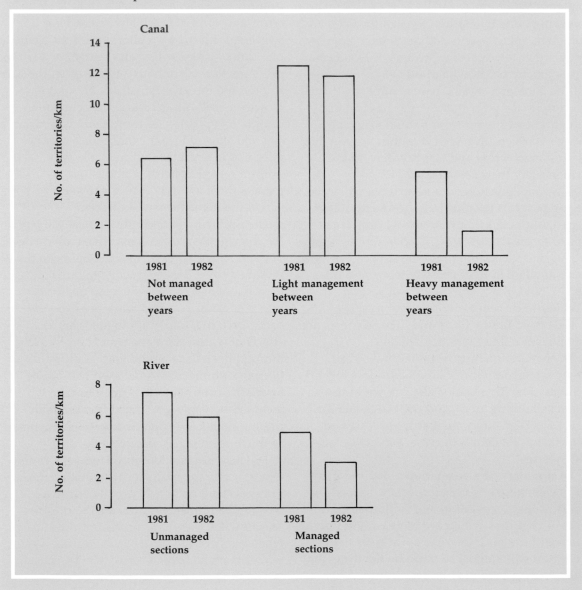

114

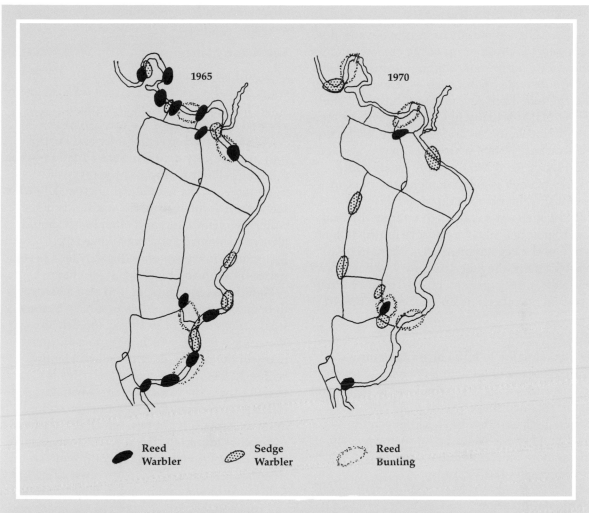

1965   1970

| Reed Warbler | Sedge Warbler | Reed Bunting |

was the destruction of nearly all the emergent and bankside vegetation. As a simple way of alleviating some of the destruction, Taylor advocated working only from one side of the channel. This should ensure that some vegetation is left, both in the water and on the bank, which in turn will aid the retention of riparian birds and should also help to reduce bank erosion.

Plot B: Similar results were obtained from a farm in Dorset[7]. A river bordering the farm CBC plot was dredged and straightened during 1967, work which also involved the removal of much of the bankside vegetation. The maps illustrate the distribution of territories of Reed Warbler,

Sedge Warbler and Reed Bunting in 1965 (before the management work) and in 1970 (3 years after it). The numbers of Reed Warblers were considerably reduced and all the Sedge Warblers and Reed Buntings redistributed territories from the riverside into nearby hedges and ditches. Subsequent to 1970 all three species virtually disappeared from the farm altogether following further management work (data in CBC files), although there was also a widespread decline of the species for other reasons[8,9].

Sources: Taylor[6], Williamson[7].

Dredging

This is designed mainly to improve water flow by a general cleaning up of the channel and its immediate surrounds, although it may involve more substantial work such as straightening of the channel. Dredging removes most, if not all, of the plants actually in the water, but the reeds and other plants can and should be left in any bays or backwaters which may exist. A Sedge Warbler or Reed Bunting only needs a few metres of such vegetation along the river to establish a territory, as do Moorhens.

Most such work is carried out from the bank and therefore is likely to involve destruction of much of the vegetation as well. This can be at least partially alleviated by only working from one side and leaving any bushes, trees and long grass on the other side untouched (see Box 10.3).

Straightening

The main value to birds of straightening a watercourse will lie with any backwaters and oxbows which are left. With vertical banks and water retained these will be ideal for Kingfishers (see Box 10.2). Moreover, the likelihood of the water being shallower will also mean more emergent aquatic vegetation which is also much used by birds.

Bank stabilization

Ideally, the banks of a waterway will be left untouched with some vegetation on them. Management work, however, very often leaves them bare and this increases the potential for erosion, which needs to be prevented. A fairly cheap but rather unsightly means on canals and larger rivers is to put in interlocking steel piles or solid concrete revetments. Neither of these will permit plant growth nor do they provide any irregularities for perches or nests. Uncemented natural stone provides some ledges which may be used as nest sites by Pied or Grey Wagtails, and gabions (wire boxes filled with rubble) or one of the geotextiles now available will permit plants to establish. Consequently both hide any 'unnatural' structures and provide habitat for birds and other wildlife. More specific details are available in Chaplin's recent book[10].

Vegetation Management

In most cases some vegetation can be left in the water and alongside the channel with no loss of water flow. Any necessary clearing of vegetation should be done mechanically.

On unmanaged stretches trees and bushes may become very dense. In such situations some thinning is likely to improve the range of vegetation and the water quality especially when the trees or shrubs are on the south side of the waterway. Trimmings and other brash should always be removed and not allowed to accumulate in the water as this may lead to some eutrophication and blocking of the channel.

Planting trees or bushes will also add variety to open waterways but this is subject to some legal constraints. For example, the British Waterways Board, which is responsible for many of the canals, does not allow planting within 12 feet of the edge. Mainly this is because the roots can cause considerable damage to the banks, as can any burrowing animals, especially rabbits, which are attracted.

Another important factor in the management of bankside vegetation is the control of livestock. Cattle and sheep are often grazed in fields next to watercourses but their trampling can cause major damage both to the banks and the vegetation. In many cases a fence is needed to prevent access to the water except at carefully defined points. Similarly, problems may arise with erosion if the adjacent land is ploughed right up to the edge of the watercourse. A strip of rough grass with or without trees or bushes is preferable.

Disposal of spoil

In some cases the mud and other spoil from any channel which is dredged out is used to build a necessary flood barrier. Otherwise it is often simply spread out on top of the bank, which usually restricts plant growth on the bank or, worse, it is dumped in the nearest damp hollow. These hollows are usually valuable for conservation themselves, including birds, especially if they are in wet fields. Disposal elsewhere is better.

Other special measures

Other particular measures may help some birds although their incorporation may require specific permission from the relevant authority.

Explicitly putting in shallower areas of water, and particularly creating riffles over stones, rocks or concrete can create feeding habitat for specialist species such as Common Sandpipers and wagtails.

A few species will use nestboxes placed under bridges. Dippers and wagtails (Grey and Pied) are potential colonists and the best boxes for these are of the open fronted variety, rather than the traditional type as used for tits in woodland with a small hole in the front. The BTO Nestbox Guide[11] gives details.

10.4　Summary and Management Recommendations

General Importance

1 Rivers and streams are especially important for Mallard, Yellow Wagtail and Reed Bunting though others, including a whole range that are not typical farmland species, will use them. In the uplands Dipper, Grey Wagtail and Common Sandpiper are almost confined to faster flowing streams.

Physical Features

2 Width, depth and rate of flow have little direct effect on birds but have major influence via the amount of vegetation in the water.

3 Water quality affects some species via food supplies. Kingfishers are especially susceptible to this in the lowlands. As in ponds, eutrophication, caused by runoff of fertilizers or by an accumulation of cut vegetation, leads to less life in the stream and fewer birds.

Vegetation

4 Vegetation in the water provides food for some species and shelter and cover for all species which use the waterway. The species of plant does not seem to be especially important.

5 The strip alongside the waterway acts as a corridor of often less intensively farmed land, and as such will attract a variety of birds. Rough grass alongside will be occupied by Sedge Warblers, Reed Buntings and others and perhaps Barn Owls will use them for hunting. Trees especially pollards with holes will attract Little and Tawny Owls among others.

Management Practices

6 Management of waterways should be directed towards creating diversity of structure. Vegetation should be retained wherever possible, although drainage considerations may dictate some removal.

7 Some rough grass, with or without trees, will enhance the landscape value of a river as well as increasing bird numbers.

8 Working from only one bank, rather than both, is preferable because of retaining vegetation.

9 Backwaters and oxbows may be especially useful for birds such as Kingfishers in otherwise bare straight watercourses.

10 For bank stabilization it is preferable to use geotextiles or uncemented natural stone rather than interlocking steel plates or concrete.

11 Some vegetation removal may be needed to reduce excessive shading. Such work should be avoided in the birds' breeding season and when berries are ripe in the autumn.

12 Vegetation should preferably be reduced mechanically rather than chemically. Herbicides are often especially toxic in aquatic environments and can affect long stretches of waterway downstream.

13 Livestock should only be permitted access at certain points, not along the whole length of the field.

14 Spoil from any dredging should not be dumped in damp hollows which are useful for conservation themselves.

15 Nestboxes can be placed under bridges for such as wagtails, and shallower areas or riffles can be created on some streams to increase feeding areas.

References

1 Marchant, J.H. & Hyde, P.A. 1980. Aspects of the distribution of riparian birds on waterways in Britain and Ireland. *Bird Study* 27: 183–202.

2 Ormerod, S.J. & Tyler, S.J. 1987. Dippers (*Cinclus cinclus*) and Grey Wagtails (*Motacilla cinerea*) as indicators of stream acidity in upland Wales. *ICBP Technical Publication* 6: 191–208.

3 Smith, K.W. 1983. The status and distribution of waders breeding on wet lowland grasslands in England and Wales. *Bird Study* 30: 177–192.

4 O'Brien, M.R. in preparation and personal communication.

5 Shawyer, C. 1987. *The Barn Owl in the British Isles. Its Past, Present and Future.* The Hawk Trust, London.

6 Taylor, K. 1984. The influence of watercourse management on Moorhen breeding biology. *Brit. Birds* 77: 141–148.

7 Williamson, K. 1971. A bird census study of a Dorset dairy farm. *Bird Study* 18: 80–96.

8 O'Connor, R.J. & Shrubb, M. 1986. *Farming and Birds.* University Press, Cambridge.

9 Marchant, J.H., Hudson, R., Carter, S.P. & Whittington, P.A. 1990. *Population Trends in British Breeding Birds.* British Trust for Ornithology, Tring.

10 Chaplin, P.H. 1989. *Waterway Conservation.* Whittet Books, London.

11 du Feu, C. 1989. *Nestboxes.* Field Guide no. 20. British Trust for Ornithology, Tring.

CHAPTER 11

Farm Buildings

Farmhouses and other farm buildings are a conspicuous feature of most farmland in Britain. The buildings have several points of interest in themselves, but the immediate surrounds are also important as they are often rather different from the wider farming countryside — small oases in fact. Many farmhouses have a garden, most have a farmyard nearby, there are often some trees, and there may be a pond or a stream nearby. All these contribute to the diversity of the area and, with the exception perhaps of the farmhouse itself, all are potentially amenable to management for birds and other wildlife.

11.1   The importance of farm buildings

The area around farm buildings may hold a high proportion of the birds on a farm. For example, the 2% of a Sussex farm which comprised the farmhouse and its associated grounds held 46% of the hedgerow and garden birds breeding there[1]. This was partly because that area also held 20% of the hedge line on that farm.

Another study, encompassing a whole parish of 537 ha in Cambridgeshire[2], found that the total bird density was 7.5 times greater in the village area (70 ha) compared to the farmland area (467 ha). Half of the territories in the village area were associated directly with buildings and their immediate surrounds, and the majority of these were in the gardens of the older houses.

On farms, there are a few species, including House Sparrow, Starling, Collared Dove and Swallow, which nest mainly in or on the buildings themselves, and seldom occur away from them (see below). However, it appears that some birds not directly associated with the

buildings are also attracted, whilst there are others which actively avoid the vicinity of buildings. Some figures from CBC plots are in Box 11.1. Birds associated with woodland and hedgerows seem to be either attracted or unaffected, but field nesting species seem to avoid the vicinity of buildings probably because of the disturbance. Increased disturbance may also be a reason for some other (often larger) species avoiding buildings and their surrounds, and it certainly causes problems for some Barn Owls (see below).

Similar factors seem to affect the same birds at other seasons. In winter many more hedgerow and other small birds are found near to farmyards than in hedges away from buildings, and field species such as Lapwings, Golden Plovers and some thrushes seem to be more common in fields away from buildings[4,5]. The only quantitative study in winter showed that, for explaining the numbers of hedgerow birds on a 5 ha study plot[6], the area of gardens in the surrounding 2.5 km^2 was the most significant of a wide variety of habitat characteristics measured.

Clearly, therefore, the area around farm buildings is attractive to many birds, although the higher level of disturbance seems to deter at least some species nesting or feeding mainly in the fields.

11.2   The effects of particular features and their management

Farmhouses

The majority of farmhouses are at least a few decades old and some are several centuries old. Many of them have nooks and crannies, especially in the roof, which may be used as nesting or roosting sites by hole-nesting birds. However, a farmer or indeed any other house owner is perhaps less likely to allow birds to roost or nest on his house, and therefore create opportunities for them, than on any outbuildings. Many of the suggestions for managing or adapting sites to help birds are the same for both houses and outbuildings, and as there are more possibilities for outbuildings most of these suggestions are considered in the next section.

For two bird species, however, houses are more suitable than outbuildings for nest sites. These are the Swift and the House Martin, both of which are summer visitors to Europe and nest most commonly on buildings. Some specific details of what these birds, and the Swallow, require are in Box 11.2. House Sparrows, Starlings and Jackdaws also sometimes nest in holes under the eaves or in chimneys but again are more common in outbuildings. In addition several species have been recorded nesting in thick creepers on house walls, with Blackbird, Song Thrush, Robin and Chaffinch perhaps the commonest. These and others will also roost in such sites in winter.

In summary Swifts and House Martins are two species which can be attracted to nest on houses although others may use holes in the roof, chimneys or any creepers.

Box 11.1 The numbers of different groups of birds around farmyards in the breeding season

The density of birds in the vicinity of farm buildings (within a 100 m circle) was compared to that along hedgerows (100 m circles centres on randomly chosen points along hedges on the same farms). Density both per hectare and per 100 m length of hedge were estimated, the latter to take account of the usually greater density of hedges around farm buildings. The diagram shows the results for six species, two of each of three groups. Firstly, those species which can be considered to be of woodland origin, represented here by the Blackbird and Chaffinch, are seen to be much commoner near buildings (the B column) than away from them (the F column), even allowing for the length of hedge. Others in this group include Wren, Robin, Blue Tit and Great Tit. Secondly, field nesting species, such as Skylark and Lapwing, were commoner away from farm buildings. And thirdly, those species associated more with hedgerows and scrub than woodland, such as Yellowhammer and Linnet, were found in similar numbers in both areas.

A similar result was found for field nesting species in the Netherlands[3]. Breeding Lapwings and Black-tailed Godwits avoided the vicinity of buildings and the avoidance extended further from an active farmyard than from a lone house or other building.

Sources: CBC data, van der Zande et al.[3].

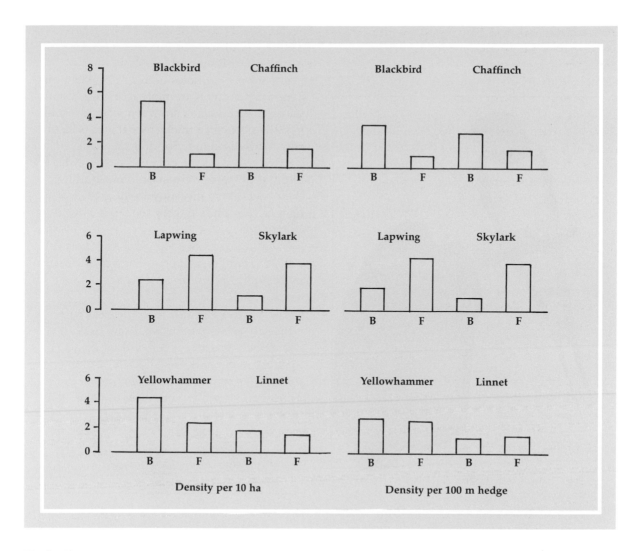

Density per 10 ha   Density per 100 m hedge

Outbuildings and farmyards

Outbuildings of various kinds are a feature of all farms and some contain livestock, hay, straw, grain or machinery. In almost all cases there is also an associated yard, usually with a hard base, an accessible water supply and very often some odd corners of waste ground. The house and any other buildings provide shelter, and the presence of a house and often livestock may well make the general area of the farmyard a few degrees warmer than the surrounding farmland. This could be especially important in very severe weather. A water supply can be very important as well, again especially in cold or very dry weather.

Outbuildings and farmyards are used by a variety of birds. Some of these are beneficial to the farmer, some are neutral and some are a nuisance. Among the first two groups are Swallows and especially Barn Owls, and chief among the latter are probably House Sparrows, Starlings and Collared Doves. The needs of Swallows are discussed in Box 11.2, and those of the Barn Owl in Box 11.3, and the latter box also discusses some more general aspects of the conservation of this species.

As noted in Box 11.2, Swallows need a permanently open entrance. The problem is that this can also let in other, often less desirable, species especially if there is grain or potential food available in the building. House Sparrows, Starlings and Collared Doves can be pests in

Box 11.2 Buildings as nest sites for Swift, House Martin and Swallow

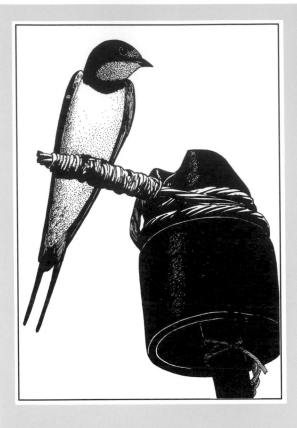

especially in the apex of a gable[8]. There is usually some shelter from the elements and both this species and the Swift need a clear route to be able to fly in directly. Both species will use artificial nestboxes[7] and, although House Martins prefer to build their own nests, the presence of artificial ones may stimulate them to start a new colony.

Swallows require a very different sort of site. A typical nest is on a ledge inside an open building. An essential is a continuously open entrance of at least about 50 cm wide and 15 cm high although, when prospecting, they will normally only enter where a door or an opening of equivalent size has been left open. They will use almost any ledge or projection within the building to place their nest, (even a nail!), but they prefer to be near the roof and close to the exit, probably to be able to keep clear of potential predators such as cats and rats.

Swifts and House Martins only rarely build their nests below about 5 m from the ground, and hence only rarely nest on bungalows or other single storey buildings. Both species usually nest in or near the roof. Swallows on the other hand prefer to nest inside a building on a ledge and therefore very rarely use houses, although barns, sheds and lean-to structures against a house may be used.

For nest sites Swifts choose a hole, either in the roof or just under the eaves, with a platform of some kind behind it on which to lay their eggs. They do not really build a nest as such, but simply put their eggs on any flattish surface. House Martins build a mud nest and the favourite site is the junction between the wall and the eaves,

In some areas at least Swallows are more common when livestock are present[9]. This is probably because of both increased warmth due to the animals and the greater insect food supply attracted by them.

Both Swallows and House Martins build their nests primarily out of mud, so if there is no natural source available nearby, the provision of a muddy puddle will aid colonisation.

Sources: BTO Nestbox Guide[7] and others

some situations, mainly through their fouling of stored products or machinery. All three may feed, nest or roost in buildings although the Collared Dove will only use very open ones as nest sites. The only practical solution is to prevent all birds from entering buildings with such stored products, and only open to Swallows those buildings where there are no worries of fouling. Swallows will nest in quite close proximity to each other if numbers are high and potential sites scarce, but the species has been declining in recent years[11], and large colonies are now uncommon.

House Sparrows, Starlings and sometimes Collared Doves utilise outbuildings as nest sites, as well as for feeding. The first two nest in holes, especially in or near the roof, and both species being semi-colonial only compounds any problems. The only certain way to keep them out is to block all suitable holes, but this is difficult as the Starling will nest in chimneys and House Sparrows in small cavities near guttering. In both these latter situations nests may cause blockages and therefore further potential problems.

All these species also feed in and around farmyards at all times of year, especially the House Sparrow. This is predominantly a seed eater all the year, but will eat other things including scraps and some insects. In the breeding season the Starling eats predominantly insects and other invertebrates and as such is often beneficial, but even at other seasons, when it turns much more to fruits, seeds, scraps and anything else it can find, it only rarely causes any problems unless it invades livestock feeding troughs. This is most common during and soon after severe weather when other food is in short supply. The Collared Dove is almost exclusively a seed eater all the year.

In winter several other species, especially of seed eaters, are attracted to farmyards. Yellowhammers in particular are attracted to yards in which cattle are housed for the winter[12]. Birds come in to feed both on spilt cattle food and on any other seeds they can find. Like Starlings, Yellowhammers will eat the food direct from the troughs even with the cattle present, but there are rarely sufficient numbers to cause problems.

Water is the other main attraction of farmyards especially in freezing or very dry weather. Again it is seed eaters which are the main species attracted as, unlike insectivores, most seed-eaters cannot get sufficient moisture from their food. A puddle or bird bath will be a great attraction and it should preferably be sited such that there are safe retreats nearby but which cannot harbour prowling cats.

The majority of farmyards have some odd corners. Often they are partially overgrown with herbs and shrubs, and may contain anything from old rusty machinery to piles of wood, old straw or rubbish. Any such area provides cover which can be used as a nesting or roosting site by birds, especially if it is a corner which is relatively undisturbed. Even very small patches may be used by for example nesting Robins or Wrens.

A few plant species typical of waste ground and similar corners are especially valued as food by some bird species. For example thistles and especially teasel will attract Goldfinches, and brambles attract several species to eat the fruits. In addition stinging nettles are a favoured food plant for several common species of butterfly such as the Small Tortoiseshell and the Peacock.

These areas may be important sites for birds

around the farmyard and should not be 'cleaned up' too intensively especially during the nesting season or when the seeds are available. Some cleaning up during the winter may, however, be necessary as such areas can also harbour rodents.

In summary outbuildings and farmyards provide suitable nest sites for several hole nesting species and feeding areas for seed-eaters. Water and corners of waste ground may also be valuable.

Gardens

Birds are attracted to gardens particularly the larger and more mature ones. This is especially so in winter and the attraction is enhanced for many birds if food or water is provided. Gardens and gardening for birds have been dealt with more fully in other books (e.g. ref. 13), so the subject will not be covered in great detail here.

There is nothing special about farmhouse gardens in comparison with others except that they are usually surrounded by open country rather than by other gardens. Therefore a farmstead garden containing shrubs or trees will be particular attractive to birds which prefer such vegetation. Many birds which are common to hedgerows, such as Blackbirds, Chaffinches and Blue and Great Tits, occur regularly in gardens, and in winter the numbers of birds can be increased considerably by providing food on a bird table. Scraps, seeds and peanuts all attract birds especially in severe conditions, and a pond, or bird bath, which birds can drink from or bathe in, is also a great attraction in dry or very cold weather.

Most of the less usual garden birds are normally associated with woodland, e.g. Nuthatch and Great Spotted Woodpecker. They will be rare in farm gardens unless there are some areas of woodland or scrub nearby. Open field species are always rare in gardens and some other typical farm hedgerow or scrub species also occur only rarely, e.g. Linnet in 2.2% and Yellowhammer in 6.4%[13]. The latter species is interesting as it is often in farmyards in winter (see above), and it may occur more commonly in farmhouse gardens than the figures for gardens overall suggest. A similar situation may exist with Reed Buntings, which are mainly cold weather specialists in gardens in general.

Gardens have several particular features which attract birds. Trees and shrubs may be used as nest sites and some produce berries, the latter including typical native hedgerow shrubs and, very often, a variety of other species both native and exotic. An orchard in or adjacent to a garden will also bring in birds to eat the fruits, both on the trees and on the ground. The most abundant of the birds will probably be typical garden birds such as Blackbirds and Starlings, but others more typical of fields, e.g. Redwings and Fieldfares, will also often arrive in hard weather. Finches and even insectivores may be attracted as well, the latter both by the fruit itself and by insects attracted by any that are over ripe.

Lawns provide open ground where it is easy for the birds, especially thrushes, to get around and find earthworms and other invertebrates. Dunnocks and Robins prefer areas which are slightly less open, so that flowerbeds and vegetable patches provide ideal feeding sites for these species. The Robin is particularly attracted to active working and newly turned soil.

Most of the suggestions for the management of gardens specifically for birds apply as much to farmhouse gardens as to any other (see ref. 13). Planting trees as song and observation posts, planting particular species of shrubs for nesting and roosting cover or as a berry source, provision of water for drinking and bathing, and specific provision of food (scraps, nuts, fruits, seeds) have all been mentioned. One can also provide one or more nestboxes for hole nesters to nest or roost in. Both the typical sort for tits, and the open-fronted type which may be used by Robins or Spotted Flycatchers, are often useful additions especially if there are few natural holes — see also Box 8.5 and ref. 7).

In summary farmhouse gardens may be an island of woody vegetation and hence attract some birds but unless there is some woodland nearby many of the scarcer garden birds are

unlikely to occur. Planting of berry-bearing shrubs and provision of food, water and nestboxes can all enhance the numbers of birds present.

Other features

There are sometimes other features near a farmyard which may enhance the attractiveness of the general area to birds. These include groups of trees, even small woods or plantations, a more substantial pond than a garden can provide or a stream. The last occur in particular near older houses. These features are all considered in detail in earlier chapters and the majority of the suggestions and management practices apply as much to features near to buildings as to those anywhere else on the farm. One different aspect is that areas near buildings are likely to be subject to greater disturbance. Hence those birds which may be adversely affected may not appear without some effort to reduce disturbance. Specific measures may only be possible around larger complexes of buildings, but for example routing regularly used tracks around only one side of a building or group of trees, planting a screen of shrubs between such a track or farmyard and a pond, and keeping hedges linking the farmhouse area with other features will all help to reduce the possible effects.

11.3 Summary and Management Recommendations

Importance

1 Farmhouses and their immediate surrounds can hold a large proportion of the birds on a farm, especially of those which occur primarily in woody vegetation, e.g. Blackbird and Chaffinch. Some field-nesting species, e.g. Lapwing, avoid buildings largely because of the increased disturbance.

Buildings

2 Few birds will nest on houses although Swifts and House Martins may nest in the roof or under the eaves respectively. House Sparrows and Starlings may nest in chimneys or other holes and a few birds will nest or roost in thick creepers.

3 Swallows often nest in outbuildings as long as there is a permanent opening for access, and they are tolerant of a good deal of disturbance.

4 Barn Owls might occur in more secluded corners. This species is giving rise to major conservation concern and activity — see Box 11.3.

5 Open doors and holes may, however, also attract some unwanted species such as House Sparrows or Starlings, and it is probably best to block entrances to buildings where fouling could cause problems, e.g. grain stores.

Farmyards

6 Farmyards provide food supplies to a range of bird species. Yards with livestock are particularly attractive and, although some birds may even feed from actively used feed troughs, only the Starling is likely to be a nuisance in this regard.

7 A water supply attracts many birds, particularly seed-eaters, and especially in dry and cold weather.

8 Odd 'waste' corners should be encouraged and left undisturbed during the breeding and berry and seed seasons.

Gardens

9 Gardens can provide a variety of attractive features. Trees and shrubs provide nest and roost sites. Several shrubs provide berries as food, while lawns and flowerbeds provide other feeding sites. A bird table can attract species especially in winter. A bird bath can prove even better in dry weather, and in cold weather if the ice is kept off. Nestboxes can be situated on trees or the house.

Other Features

10 Surrounding features such as small woods or ponds should be encouraged preferably leaving some areas within them relatively undisturbed by humans or machinery.

11 Links should be retained with other habitat features by means of hedges or strips of uncultivated ground.

Box 11.3 The Barn Owl and its conservation

Conservationists are becoming increasingly concerned about this species. The Hawk and Owl Trust organised a nationwide survey of its distribution, numbers and habitat requirements in the early 1980s[10], and showed that, as had been suspected, its numbers had declined by about 60% over the previous 50 years. It is an emotive bird and one which is beneficial to farmers. This is because its food is primarily rodents and they are often hunted from around farmyards. Therefore any measures which can be taken to help the Barn Owl are to be encouraged.

When nesting, the Barn Owl is very sensitive to disturbance and hence any nesting site will be in the quietest available corner. In many areas it uses barns, other outbuildings or haystacks, but in some parts of the country, especially in drier areas in the east, they prefer to nest in holes in large trees[10]. In buildings the best sites are fairly high up on a dark ledge or in a cavity, and with permanent access to the outside. This is preferably via a hole in the wall rather than a door or other entrance. They adapt readily to nest boxes, with the best and most commonly used being old tea chests or barrels. However, great care should be taken with the siting especially if birds are already present — see ref 7.

Attempts are being made to try to reverse the decline, although some of the major causes are hard to do anything about. Britain is at the northern edge of the species' range and it is subject to quite severe mortality in winter, especially in the colder ones. The major cause of the decline seems to be a reduction in the amount of feeding habitat available. This is mentioned with some suggested remedies in chapters 4 and 5 (p. 37 and p. 51).

Some well-meaning people are trying to increase the population by releasing captive-bred birds. Barn Owls are very easy to breed in captivity and there may be 400 or so schemes in operation which are releasing birds, some on quite a large scale[10].

However responsible conservationists are becoming increasingly worried by the proliferation of these schemes, especially with respect to the effect of the introduced birds on any wild birds which are already present. Wild birds are notoriously difficult to detect even if they are regularly using a building as a roosting or nesting site. An introduced bird or birds could have a major disrupting effect, especially as the decline in many places is because of lack of food resources and hunting areas rather than any lack of roosting and nesting sites. Obviously, if there is little or no food available any released birds are not going to survive in the longer term and certainly not breed. The little evidence there is suggests that most released birds do not survive for very long after any initial period of artificial feeding. If birds are released into an area where wild birds still occur but there is little food, then the competition could lead to the demise of the wild birds too.

In the near future the Department of the Environment intends to place the Barn Owl on Schedule 9 of the Wildlife and Countryside Act 1981 which will require this activity to be licensed.

Sources: Shawyer[10] and others.

References

1 Shrubb, M. 1970. Birds and farming today. *Bird Study* 17: 123–144.

2 Wyllie, I. 1976. The bird community of an English parish. *Bird Study* 23: 39–50.

3 van der Zande, A.N., ter Kewis, W.J. & van der Weijden, W.J. 1980. The impact of roads on the densities of four bird species in an open field habitat — evidence of a long distance effect. *Biol. Conserv.* 18: 299–321.

4 Spencer, R. 1982. Birds in winter — an outline. *Bird Study* 29: 169–182.

5 R. Spencer personal communication and personal observations.

6 Arnold, G.W. 1983. The influence of ditch and hedgerow structure, length of hedgerows, and area of woodland and garden on bird numbers on farmland. *J. appl. Ecol.* 20: 731–750.

7 du Feu, C.R. 1989. *Nestboxes*. Field Guide no. 20. British Trust for Ornithology, Tring.

8 Bell, C. 1983. Factors influencing nest-site selection in House Martins. *Bird Study* 30: 233–237.

9 Møller, A.P. 1983. Breeding habitat selection in the Swallow *Hirundo rustica*. *Bird Study* 30: 134–142.

10 Shawyer, C.R. 1987. *The Barn Owl in the British Isles. Its Past, Present and Future*. The Hawk Trust, London.

11 Marchant, J.H., Hudson, R., Carter, S.P. & Whittington, P.A. 1990. *Population Trends in British Breeding Birds*. British Trust for Ornithology, Tring.

12 O'Connor, R.J. & Shrubb, M. 1986. *Farming and Birds*. University Press, Cambridge.

13 Glue, D.E. 1982. *The Garden Bird Book*. Macmillan, London.

CHAPTER 12

General Conclusions and Future Prospects

Previous chapters of the book have covered the management of different habitat features on farms, and described how these can be made more suitable for birds, as well as other wildlife. Specific suggestions have been made for benefitting particular bird species, and many of these suggestions are not very different from what are already common, even standard, practices. Hence, often they involve no major costs to the farmer.

The first part of this chapter sets out eight general principles for benefitting birds in farmland, and therefore serves, in effect, as an overall summary and generalisation of the specific details of earlier chapters.

The second part of the chapter considers briefly some possible scenarios for the future of farmland and the birds which live in it. Some areas on which more research is needed are outlined.

12.1 Eight general principles for managing farmland to benefit birds

1 Concentrate conservation efforts on natural and semi-natural areas except in those few cases where species of special interest are present in the actively farmed area.

On most farms the largest numbers of birds and those of greatest interest to conservation are in the hedges (ch 3), small woods (ch 8), ponds (ch 9) and other natural or semi-natural areas rather than in the fields or other cultivated areas. There are, however, a few exceptions, for example where there are breeding waders in the fields (p. 69).

2 Particular emphasis should be placed on retaining old or ancient features, including those which have been managed in the same

traditional way over a long period.

Ancient hedges and woods are especially important for conservation as they often hold rare and interesting species and communities (see p. 91). Similarly fields which have not been improved or subjected to a change of management, can hold such species, in particular some old grassland such as lowland hay meadows (p. 71).

3 Retain, or create, as wide a range of different habitat features, and as wide a range of different structures within them, as is practicable.

There are likely to be more individuals and species of birds on farms with a greater variety of habitat both within and between features (ch 2). A diversity of structure may occur naturally, but it can also sometimes be created by management. However one must be careful not to fragment blocks of habitat too much (see especially p. 15 and p. 85).

4 Preserving or managing existing features is usually more profitable for wildlife, and more efficient economically, than creating a similar new feature.

In existing features some relevant species (plant and animal) will probably be present already. In addition, some of the more interesting and rarer species are not very efficient at dispersing so it might be a long time before a new feature is colonised.

5 For many habitats in farmland some management is essential to retain their interest and diversity.

Many habitat features in farmland are more or less derelict, and such a state is not necessarily the best for birds and other wildlife (see especially ch 3 (hedges) and ch 8 (woods)).

6 Ensure that management of areas does not destroy existing wildlife that may be of

greater interest than species attracted by the change.

Birds and other wildlife on a site may be destroyed or be forced to move elsewhere, at least temporarily, by habitat management. It is important to consider whether such species are of greater interest than those encouraged to colonise by any changes (see for example p. 30).

7 Ensure that management and other work only affect the intended areas and features, and do not have detrimental effects on nearby areas.

Most areas in farmland are closely interlinked with others, and management work on one site may have implications for nearby features. It is also very important to ensure that treatments such as fertilizers and pesticides do not drift or leach into areas where they will be harmful (p. 31).

8 It is essential to set clear objectives before starting any management work, and to plan and carry it out carefully, so that these objectives are met without prejudicing any other aspects of the farm.

If in doubt about what to do, it is best to take advice before any work is started. Such organisations as the Agricultural Development and Advisory Service and the Farming and Wildlife Trust are always willing to help with specific advice if asked. Addresses of these and other such organisations are in Appendix 2.

12.2   Future prospects

Since the Second World War the highest priority of farmers and farming has been the large scale production of food by the most cost-effective means possible. This objective has been encouraged by economic incentives in the form of grants, loans and guaranteed prices from the British government and, since 1973, from the European Community. It has led to most of the changes seen in farmland over the last forty to fifty years.

Most of the changes come under the heading of increased intensification. Major specific ones include drainage, removal of hedges and small woods, conversion of marginal land into arable or improved grass, greater input of fertilizers and pesticides, greater stock density, and changed timing of farming operations associated with different cropping regimes. These are discussed in relevant sections of earlier chapters.

Although farmers have the health of the land at heart, economics often limit their opportunities to do things which benefit birds and other wildlife. However, this situation is changing. Public perception of green issues has highlighted several aspects, and there are now more incentives to farm in an environmentally friendly way. These are likely to expand, along with alternatives to producing crops which are in surplus. Some of these are mentioned below, and more details can be found in books and leaflets produced by the schemes or organisations involved.

Protected Areas

Some areas are already subject to particular restrictions and guidelines on what can be done on the land. Often, these refer mainly to the landscape rather than natural history, and very few are specifically aimed at conserving bird species. However, birds are likely to benefit in most cases, because the permitted farming practices usually involve less intensive methods than in unrestricted areas. In general, management which benefits the landscape will also benefit birds.

Areas subject to some degree of protection and restriction include the National Parks, Less Favoured Areas, Environmentally Sensitive Areas, Areas of Outstanding Natural Beauty, and Sites of Special Scientific Interest. All areas currently within one or more of these schemes seem likely to continue under some form of protection, and they serve as important reservoirs for birds and other wildlife on farms.

It seems likely too that new areas will be designated into some of these categories, and there may well be other new categories designated in the future with varying degrees of protection and restrictions on what can and cannot be done.

Diversification

Some farmers have branched out into new activities. These include farm tourism and the planting of novel crops. Some can be done under one or more of the Incentive Schemes, but others are carried out as a farm business venture without any financial assistance from the government or other body.

Most arable farmers grow a so-called break crop within their cereal rotations and these are mostly mainstream crops such as oilseed rape or a temporary ley. However, other farmers always grow one or two fields of a particular unusual crop if they have identified a market, and a few concentrate almost exclusively on one or more of these, including herbs, ornamentals and other market garden products. However, none of them, with the possible exception of sunflowers, is of particular interest to birds, and there is not a sufficient market for sunflowers to make an impact such as, for example, to reverse the decline in availability of food for some of the seed eating species in farmland.

Some farmers plant trees for small scale timber production, usually planted under the Farm Woodland Scheme or other grants. However, particular species, such as cricket bat willows, may be planted in a few places for a particular market. As with other unusual crops, the areas grown are very small and are unlikely to affect birds on an important scale.

On the livestock side, a few farmers now specialise on deer farming. This involves a high initial capital outlay and is not undertaken without considerable thought but, as with most unusual crops, from a bird's viewpoint it is relatively unimportant.

Farm tourism is becoming increasingly popular, especially in coastal holiday areas. Diversifying into farm tourism typically involves a farm nature trail and may include animals, often of lesser known breeds, held in small enclosures. Obviously, nature trails are likely to benefit birds and other wildlife, and such trails are likely to be placed in those parts of a farm which already contain more and interesting species of wildlife. Birds are usually mentioned prominently in the associated information booklets, as they are very popular with the public and there is substantial background information on the habitat of most species.

Some farmers opt out of farming altogether. They may convert land into golf courses, lakes for fishing or boating, or even building plots. If the change involves creation of more natural or semi-natural habitat it will benefit birds and other wildlife.

Organic farming

The amount of land being farmed organically is increasing slowly and, because no synthetic chemicals are used, one might expect birds and other wildlife to benefit considerably. Unfortunately, very little research has been done and there are only a few indications of any effects on bird populations. One major study involving birds was carried out in Denmark[1]. This compared birds counted at points on organic farms with counts at similar points on conventionally farmed areas. Some results are given in Box 12.1, and show that land farmed organically held considerably more birds, and of more species, than land which had been farmed conventionally.

Similar results were obtained by two studies in the United States (quoted by Arden-Clarke[2]), but here the results were more difficult to interpret because the organic farms had a greater diversity of crops, and the quality of some non-crop habitat was greater than on the conventional farms studied.

Details of how an organic farm can be set up on a large scale, and some of the trials and problems encountered and overcome, can be found in Barry Wookey's book[3].

Grants for habitat creation/renovation

There is a variety of grants and loans available from several sources, including MAFF, the Countryside Commission and the Forestry

Box 12.1 Bird numbers on organic and conventional farms in Denmark

Birds were censused at fixed points on thirty one organic farms, and at matched points on farms managed conventionally.

The numbers of twenty four out of thirty five common bird species were higher on the organic farms, and for twenty of them the increase was of more than 10%. The twenty four included all the more important farmland breeding species in Denmark, such as Lapwing, Skylark, Swallow, Whitethroat, Linnet and Yellowhammer; and eleven of them were species which had declined significantly

overall in Denmark since 1976. The species also included Black-headed Gull and Woodpigeon, which breed in other habitats but feed in arable land, as well as birds more typical of small woods or gardens, such as Garden Warbler, Tree Pipit, House Sparrow and Greenfinch.

Of the eleven species which were more numerous on conventional farmland, the numbers of only five were more than 10% higher than on organic farms.

Source: Braae et al.[1].

Authority, for activities such as planting hedges, renovating ponds, and managing woodlands. Further details of current schemes can be obtained from local MAFF offices and other organisations listed in Appendix 2.

Extensification

This scheme, which has yet to be brought fully into operation, is designed to reduce by 20% the amount of cereals and beef reaching the market. This reduction target can be achieved in a variety of ways, and not necessarily by taking land out of production. Farmers are paid compensation for production lost through using less intensive farming methods.

This scheme, like the Set-Aside scheme discussed in the next section, was not designed with any environmental benefit in mind, but it could potentially have considerable benefits to birds and other wildlife.

Many of the suggestions mentioned in earlier chapters here could profitably be followed, and specific practices include reductions in pesticide and fertilizer use, leaving fields to fallow, creating flower or herb rich headlands or fields, and planting hedges or digging ponds and then managing these in environmentally friendly ways. The NCC has produced a report on all aspects of the benefit to conservation of

extensification as it applies to wildlife, and how farmers reacted to the suggestions[4]. The main conclusions as far as birds are concerned were that all management which reduces the intensity of farming is likely to benefit birds and such practices as promoting permanent grassland and spring-sown cereals, and managing hedges more sympathetically would be supported by farmers if adequate compensation were made available.

Set-Aside

A European wide Set-Aside scheme was implemented in Britain in autumn 1988, with the principal aim of reducing the cost of surplus production of cereals. Farmers were asked to take at least 20% of land growing cereals out of production completely in return for compensation payments. There were, however, some restrictions on what can be done to qualify for the payments, and the official booklets about the scheme[5] should be consulted for the details.

There are three main alternatives:

a Fallow, which involves leaving fields to fallow for either the whole five-year period (Permanent Fallow) or having different fields each year (Rotational Fallow);

b Non-Agricultural Use, which involves conversion of the land to such as small-scale industry or golf courses and which includes the creation of nature reserves;

c Woodland, which involves planting trees.

Planting woodland is usually more economic for the farmer if it is done under the Farm Woodland Scheme, many of the non-agricultural use alternatives have no benefit for birds or other wildlife, but there are several potential benefits with the Fallow Option, especially Permanent Fallow. This has also proved by far the most popular option to farmers, although it must be said that one or two of the restrictions, notably of grazing, limit the potential somewhat.

The RSPB has produced a report[6] setting out how birds can benefit from the various options under Set-Aside and this should be consulted for the details. The report concludes that Permanent Fallow is the perferred option under most circumstances. This is because it allows the creation of their 'wildlife fallow'. This will benefit Lapwings and also Stone Curlews where they occur, and in addition, putting in grassland especially into otherwise arable areas will benefit several other species. The report notes that the restrictions on grazing mean that the fallow areas are less useful than they could be for some species, for example those which require a short field layer, but this restriction is being lessened to some degree in the revised rules. It also gives specific advice on the creation of scrub, goose pasture, woodland, heathland, ponds, reedbeds, wet meadows and game crops under the Set-Aside scheme, and how these can benefit birds and other wildlife.

Countryside Premium Scheme

In five countries of eastern England the Countryside Commission has introduced a scheme whereby additional payments will be made for land included in Set-Aside which is managed specifically for the benefit of wildlife. The payments are only available for land put into the Permanent Fallow Option of Set-Aside, and the basis of what can be done are similar to some of the ideas put forward by the RSPB in their report[6]. Extra payments are available for managing Wooded Margins (including hedgerows), Meadowland (conversion of arable to grass), Wildlife Fallow (land attractive for ground nesting birds and wild flowers), Brent Goose Pasture (to create winter grazing as a means for minimising damage to crops in areas where Brent Geese are a problem), and Habitat Restoration (to restore particular wildlife-rich habitats). All of them have potential benefit for birds. Further details are available from the Countryside Commission[7].

Other Future Prospects

There will no doubt be other schemes brought into operation in the future both to deal with perceived problems and to recognise growing environmental awareness. It is to be hoped that more of them will be directed specifically at increasing the numbers and variety of birds and other wildlife which occur, or at least take due consideration of the wildlife interest.

Farmers are the chief guardians of the countryside, and they wish to help as much as they are able to within their economic constraints, and to be seen to be doing so. Equally, conservationists must acknowledge that farmers are not the unremitting destroyers of wildlife which they have been made out to be in some quarters, and that many farmers are making positive efforts to conserve the wildlife, including birds, on their farms and will continue to do so. If farmers and other landowners follow all the suggestions in this book the countryside will hold many more birds and of a greater variety in the future.

References

1 Braae, L, Nohr, H. & Petersen, B.S. 1988. *Fuglefaunaen på konventionelle og økologiske landbrug.* Miljøprojekt No. 102. Ornis Consult ApS, Copenhagen.

2 Arden-Clarke, C. 1988. *The Environmental Effects of Conventional and Organic/Biological Farming Systems.* Parts I–IV. Political Ecology Research Group, Oxford.

3 Wookey, C.B. 1987. *Rushall: The Story of an Organic Farm.* Blackwell, Oxford.

4 Centre for Rural Studies of Royal Agricultural College. 1988. *Cereal Extensification in Lowland England – an Assessment of the Benefits for Wildlife.* A report to the Nature Conservancy Council.

5 Ministry of Agriculture, Fisheries and Food & Welsh Office Agriculture Department. 1988. *Set-Aside. A Practical Guide.* MAFF/WOAG, London.

6 Osborne, P.E. 1989. *The Management of Set-Aside Land for Birds: A Practical Guide.* Royal Society for the Protection of Birds, Sandy.

7 Countryside Commission. 1989. *The Countryside Premium for Set-Aside Land.* Countryside Commission, Cambridge.

APPENDIX 1

Scientific names of animals and plants mentioned in the text. All follow current usage.

a Birds

Little Grebe *Tachybaptus ruficollis*
Great Crested Grebe *Podiceps cristatus*
Cormorant *Phalacrocorax carbo*
Grey Heron *Ardea cinerea*
Mute Swan *Cygnus olor*
Pink-footed Goose *Anser brachyrhynchus*
Greylag Goose *Anser anser*
White-fronted Goose *Anser albifrons*
Canada Goose *Branta canadensis*
Barnacle Goose *Branta leucopsis*
Brent Goose *Branta bernicla*
Wigeon *Anas penelope*
Teal *Anas crecca*
Mallard *Anas platyrhynchos*
Shoveler *Anas clypeata*
Pochard *Aythya ferina*
Tufted Duck *Aythya fuligula*
Montagu's Harrier *Circus macrourus*
Marsh Harrier *Circus aeruginosus*
Sparrowhawk *Accipiter nisus*
Kestrel *Falco tinnunculus*
Red-legged Partridge *Alectoris rufa*
Grey Partridge *Perdix perdix*
Pheasant *Phasianus colchicus*
Moorhen *Gallinula chloropus*
Coot *Fulica atra*
Corncrake *Crex crex*
Oystercatcher *Haematopus ostralegus*
Ringed Plover *Charadrius hiaticula*
Golden Plover *Pluvialis apricaria*
Lapwing *Vanellus vanellus*
Woodcock *Scolopax rusticola*
Snipe *Gallinago gallinago*
Black-tailed Godwit *Limosa limosa*
Curlew *Numenius arquata*
Common Sandpiper *Tringa hypoleucos*

Redshank *Tringa totanus*
Stone Curlew *Burhinus oedicnemus*
Black-headed Gull *Larus ribidundus*
Herring Gull *Larus argentatus*
Common Tern *Sterna hirundo*
Stock Dove *Columba oenas*
Woodpigeon *Columba palumbus*
Collared Dove *Streptopelia decaocto*
Turtle Dove *Streptopelia turtur*
Cuckoo *Cuculus canorus*
Barn Owl *Tyto alba*
Little Owl *Athene noctua*
Tawny Owl *Strix aluco*
Swift *Apus apus*
Kingfisher *Alcedo atthis*
Green Woodpecker *Picus viridis*
Great Spotted Woodpecker *Dendrocopos major*
Skylark *Alauda arvensis*
Meadow Pipit *Anthus pratensis*
Tree Pipit *Anthus trivialis*
Sand Martin *Riparia riparia*
Swallow *Hirundo rustica*
House Martin *Delichon urbica*
Meadow Pipit *Anthus pratensis*
Yellow Wagtail *Motacilla flava*
Grey Wagtail *Motacilla cinerea*
Pied Wagtail *Motacilla alba*
Dipper *Cinclus cinclus*
Wren *Troglodytes troglodytes*
Dunnock *Prunella modularis*
Nightingale *Luscinia megarhynchus*
Thrush Nightingale *Luscinia luscinia*
Robin *Erithacus rubecula*
Redstart *Phoenicurus phoenicurus*
Wheatear *Oenanthe oenanthe*
Stonechat *Saxicola torquata*
Whinchat *Saxicola rubetra*
Blackbird *Turdus merula*
Fieldfare *Turdus pilaris*
Song Thrush *Turdus philomelos*

Redwing *Turdus iliacus*
Mistle Thrush *Turdus viscivorus*
Sedge Warbler *Acrocephalus schoenobaenus*
Reed Warbler *Acrocephalus scirpaceus*
Lesser Whitethroat *Sylvia curruca*
Whitethroat *Sylvia communis*
Garden Warbler *Sylvia borin*
Blackcap *Sylvia atricapilla*
Chiffchaff *Phylloscopus collybita*
Willow Warbler *Phylloscopus trochilus*
Wood Warbler *Phylloscopus sibilatrix*
Goldcrest *Regulus regulus*
Spotted Flycatcher *Muscicapa striata*
Pied Flycatcher *Ficedula hypoleuca*
Long-tailed Tit *Aegithalos caudatus*
Blue Tit *Parus caeruleus*
Great Tit *Parus major*
Coal Tit *Parus ater*
Marsh Tit *Parus palustris*
Willow Tit *Parus montanus*
Nuthatch *Sitta europaea*
Treecreeper *Certhia familiaris*
Red-backed Shrike *Lanius collurio*
Jay *Garrulus glandarius*
Magpie *Pica pica*
Jackdaw *Corvus monedula*
Rook *Corvus frugilegus*
Carrion Crow *Corvus corone*
Starling *Sturnus vulgaris*
House Sparrow *Passer domesticus*
Tree Sparrow *Passer montanus*
Chaffinch *Fringilla coelebs*
Greenfinch *Carduelis chloris*
Goldfinch *Carduelis carduelis*
Linnet *Carduelis cannabina*
Lesser Redpoll *Carduelis flammea*
Common Crossbill *Loxia curvirostra*
Scottish Crossbill *Loxia scotica*
Bullfinch *Pyrrhula pyrrhula*
Yellowhammer *Emberiza citrinella*
Cirl Bunting *Emberiza cirlus*
Ortolan Bunting *Emberiza hortulana*
Reed Bunting *Emberiza schoeniclus*

Corn Bunting *Miliaria calandra*

b Plants

Alder *Alnus glutinosa*
Ash *Fraxinus excelsior*
Barley *Hordeum spp.*
Beech *Fagus sylvatica*
Birch *Betula pendula*
Black-grass *Alopecurus myosuroides*
Blackthorn (sloe) *Prunus spinosa*
Bracken *Pteridium aquilinum*
Bramble *Rubus spp.*
Branched Bur Reed *Sparganium erectum*
Broom *Cytisus scoparius*
Buckthorn *Rhamnus catharticus*
Buckwheat *Fagopyrum esculentum*
Bullrush *Typha latifolia*
Canary Seed (Grass) *Phalaris canariensis*
Caraway *Carum carvi*
Carrot *Daucus carota*
Cleavers *Galium aparine*
Clover *Trifolium spp.*
Common Reed *Phragmites communis*
Crab Apple *Malus sylvestris*
Dogwood *Cornus sanguinea*
Duckweeds *Lemnaceae*
Elder *Sambucus nigra*
Elm *Ulmus spp.*
European Larch *Larix decidua*
Field Beans *Vicia faba*
Field Maple *Acer campestre*
Gorse *Ulex europaeus*
Hawthorn *Crataegus monogyna*
Hazel *Corylus avellana*
Holly *Ilex aquifolia*
Honeysuckle *Lonicera spp.*
Hornwort *Ceratophyllum demersum*
Ivy *Hedera helix*
Legume *Leguminosae*
Maize *Zea mays*
Marrowstem Kale *Brassica spp.*
Mistletoe *Viscum album*
Norway Spruce *Picia abies*

Oak *Quercus robur*
Oilseed Rape *Brassica napus*
Old Man's Beard *Clematis vitalba*
Pea *Pisum satirum*
Perfoliate Honeysuckle *Lonicera caprifolium*
Pondweed *Potomageton spp.*
Poplar *Populus spp.*
Rose *Rosa spp.*
Rowan *Sorbus aucuparia*
Rye Grass *Lolium perenne*
Scots Pine *Pinus sylvestris*
Sessile Oak *Quercus petraea*
Snowberry *Symphoricarpos rivularis*
Spindle *Euonymus europaeus*
Stinging nettle *Urtica dioica*
Sugarbeet *Beta vulgaris*
Sunflower *Helianthus annuus*

Sweet Chestnut *Castanea sativa*
Sycamore *Acer pseudoplatanus*
Teasel *Dipsacus fullonum*
Thistles *Carduus spp.* and *Cirsium spp.*
Water Lilies *Nymphaceae*
Water Milfoil *Myriophyllum spp.*
Water Starwort *Callitriche stagnalis*
Wayfaring Tree *Viburnum lantana*
Western Hemlock *Tsuga heterophylla*
Wheat *Triticum spp.*
Whitebeam *Sorbus aria*
White Bryony *Bryonia dioica*
Wild Cherry *Prunus avium*
Willow *Salix spp.*
Woody Nightshade *Solanum dulcamara*
Yellow Flag *Iris pseudacorus*
Yew *Taxus baccata*

Addresses of Farming and Countryside Organisations referred to in the text, or which provide advice on managing habitat feaures for wildlife.

ADAS Food, Farming, Land and Leisure, ADAS HQ, Oxford Spires Business Park, The Boulevard, Langford Lane, Kidlington, Oxford OX5 1NZ

British Association for Shooting and Conservation, Marford Mill, Rossett, Wrexham, Clwyd LL12 0HL.

British Trust for Conservation Volunteers, 36 St Mary's Street, Wallingford, Oxfordshire OX10 0EU.

British Trust for Ornithology, National Centre for Ornithology, The Nunnery, Nunnery Place, Thetford, Norfolk IP24 2PU.

Central Science Laboratory, MAFF, London Road, Slough, Berkshire SL3 7MJ.

Council for the Protection of Rural England, Warwick House, 25 Buckingham Palace Road, London SW1W 0PP.

Countryside Commission, John Dower House, Crescent Place, Cheltenham, Gloucestershire GL50 3RA.

Countryside Council for Wales, Plas Penrhos, Ffordd Penrhos, Bangor, Gwynedd LL57 2LQ.

English Nature, Northminster House, Northminster, Peterborough PE1 1UA.

Farming and Wildlife Trust, National Agricultural Centre, Stoneleigh, Kenilworth, Warwickshire CV8 2RX.

Forestry Authority, 231 Corstorphine Road, Edinburgh EH12 7AT.

Forest Enterprise, 231 Corstorphine Road, Edinburgh EH12 7AT.

Game Conservancy Trust, Burgate Manor, Fordingbridge, Hampshire SP6 1EF.

Joint Nature Conservation Committee, Monkstone House, City Road, Peterborough PE1 1JY.

Ministry of Agriculture, Fisheries and Food Whitehall Place, London SW1A 2HH. (See ADAS above)

National Farmers' Union, Agriculture House, Knightsbridge, London SW1X 7NJ.

National Trust, 36 Queen Anne's Gate, London SW1H 0AS.

Royal Agricultural Society of England, National Agricultural Centre, Stoneleigh, Kenilworth, Warwickshire CV8 2RX.

Royal Society for Nature Conservation, The Green, Nettleham, Lincoln LN2 2NR.

Royal Society for the Protection of Birds, The Lodge, Sandy, Bedfordshire SG19 2DL.

Scottish National Heritage, 12 Hope Terrace, Edinburgh EH9 2AS.

Scottish Office Agriculture and Fisheries Department, Pentland House, 47 Robb's Loan, Edinburgh EH14 1TW.

Wildfowl and Wetlands Trust, Slimbridge, Gloucestershire GL2 7BT.

Species Index